设计原创 No.2 风景 公共建筑 城市地方 孔少凯编撰　The Designs for Poetic
Dwelling No.2 Landscape Institution & Urban Places　　　　edited by Kong Shaokai

孔少凯 *建筑体验* *设计原创* 甲骨金文书名题字

前言

广义建筑-人的安居地方,涵盖建筑、风景、城市和国土环境。建筑是诗意人生的体验,基于艺术原创、科学思维和哲学运筹,安居地方打造可以解读为一个理念、造型、技艺与系统构成过程。

图集《设计原创》与笔记《建筑体验》,是作者36年探索与发现图形文字信息积淀:以赤子之心、诗的语言和动态三维图式思维网络,抽提与开拓自然与建筑地方潜在美学特征,创造高下曲折行云流水空间与富有雕塑感的造型,提炼材料、质地、色彩与光影的纯、反差与融合,咫尺之间铸成袖里乾坤,展示五彩缤纷诗意人生与大千世界。

人创造建筑,建筑孕育人。建筑、风景、城市和国土环境是无言的诗、无声的音乐、无字的碑:聚会与交流、陈列与展示,对人的才华、素质和情趣的孕育,有着无处不在无时不在潜移默化的影响。只有上上下下、里里外外、左左右右、前前后后精心解读安居地方的字里行间,才能体验其中作为原创构思理性演绎过程的诗:

美的建筑是内涵与外形天衣无缝的吻合,首先使人眼睛为之一亮,然后引发基于理性思维的审美激动,给人以终身难忘的经历。

面对全球气候变暖生死存亡的时刻,人类必须珍惜自然水土绿地作为生命之源调节气候的天赋功能。减缓全球气候变暖的一个重要途径应当是城市的科学发展与绿色建设:

保护自然,保护历史,珍惜每一寸水土地,在有限的土地上创造建筑、风景和城市建设精品。

大学的主要责任,在于继承对详细事实和抽象结论热烈兴趣的传统,作为一种文化遗产广泛传播于世。中国大学多学科之间交流与融合刚刚起步,21世纪中国建筑学及视觉艺术相关领域理念与实践,应当成为法律、经济、管理、社会与文理各学科的相关课程,成为家喻户晓的大众智慧。愿君如海绵一般汲取知识、智慧与普世价值。

Preface

Architecture could be understood as a poetic life' experience, dwelling place could be dissected as a working process from concept, through technique & shaping, up to systematic generation, which is based on artistic creation, scientific thinking and philosophy devising strategies.

Pictures collection *The Designs for Poetic Dwelling* & notes collection *Experiencing Architecture Designs*, with poetic language and dynamic 3D figure analysis wet, abstract and open up places' Genius Loci, create stereoscopic spaces and engraved figures, refine a purity, contrast and harmony of material, texture, color and light & shadow, and produce a microcosm which is displaying a colorful poem life and Great Chiliocosmos.

Man creates architecture, and architecture pregnant people. Poetic architecture, landscape, city and territory are a sentenceless poem, a wordless tablet and a soundless music. Meeting & exchange, display & exhibition, exercise a great influence to people's talent, quality, disposition and taste.

Poetic place is a coincide between its exterior figure and interior meaning, at first making people's eyes shining, then educing people's aesthetic excitement, and finally giving people an unforgettable experience.

Facing a critical time of earth's climate warming, mankind have to preserve every inch natural water & soil land, and treasure its inborn functions as the life origin and as the inborn conditioner of earth climate, through creating the master pieces on architecture, landscape & urban design.

目录 contents

风景游憩地方
Landscape Sightseeing Places P1

P3　松江方塔园北大门 + 餐厅建筑与风景设计 The North Entrance Area Design of Pagoda Park in Shanghai 1981-1985

P7　杭州西湖国宾馆总体规划 The Master Plan of West Lake National Guesthouse in Hangzhou 1985

P9　大连金石滩风景区总体规划 The Master Plan of Jinshitan Bay Scenery Area in Dalian 1985

P11　山西北武当山风景区总体规划 The Master Plan of North Wudang Mountain Scenery Area in Shanxi 1986

P15　武陵源国家公园总体规划 The Master Plan of Wulingyuan National Park 1987

P21　绍兴东湖石宕稷寿楼古建筑设计 The Pavilion Design in East Lake in Shaoxing 1985

P23　无锡太湖大浮山度假村总体规划与建筑设计 The Design of Dafu Hill Resort in Taihu Lake 1997

P29　江苏启东大酒店建筑设计 The Design of Qidong Hotel in Jiangsu 1998-1999

P31　南京牛首山风景区总体规划 The Master Plan of Niushou Mountain Scenery Area in Nanjing 2000

P33　启东鼋头角海滨度假村建筑设计 The Design of Yuantou Cape Seashore Club 2002-2003

P39　复旦大学江湾校园风景原创构思 The Design Concept of Jiangwan Bay Campus Landscape Development of Fudan University 2005

公共建筑地方
Public Building Places P41

P43　乌鲁木齐中药厂制剂车间建筑设计 The Design of Medicine Factory in Wurumuqi 1978

P45　泉州图书馆全国获奖建筑设计 An Award-winning Scheme Design for Quanzhou City Library National Design Competition 1983

P47　太原大剧院全国邀请赛建筑设计方案 A Scheme Design for Taiyuan City Theatre National Design Competition 1984

P49　绍兴烟箩洞宾馆 + 商业展示中心 The Design of Yanluodong Hotel & Shopping Centre in Shaoxing 1985-1986

P51　上海海友花园光大会展中心建筑群设计 The Design of Everbright Buildings Complex in Shanghai 1993-1997

P57　上海第二军医大学图书馆办公主楼 + 科学实验楼设计方案 The Design of Library & Science Buildings, No.2 Army Medicine University in Shanghai 1994-1997

P61　上海照相机四厂改建设计 The Renovation Design of No.4 Camera Factory Buildings in Shanghai 1994-1995

P65　绍兴中学百年校园改建方案 The Concept Design of Shaoxing Secondary School 1994

P67　上海外滩光大银行改建设计 The Renovation Design of Everbright Bank in Shanghai Bund 1998

P83　上海爱建大厦建筑设计方案 The Scheme Design of Aijian Mansion in Shanghai 1998

P85　上海大学美术学院建筑设计方案 The Scheme Design of Art Institute, Shanghai University 1998

P93　镇江江苏理工大学教学楼建筑设计方案 The Scheme Design of Teaching Building of Jiangsu Technology University 1999

P95　上海大学宝山新校园校门区原创构思 The Concept Design of Entrance, Shanghai University 1999

P97　上海音乐学院荟思实验学校建筑设计方案 The Scheme Design of Wisdom School, Shanghai Music Academy 2001

P103　杭州萧山会展中心建筑设计方案 The Scheme Design of Xiaoshan Exhibition Centre in Hangzhou 2003

P105　山东曲阜孔子博物馆原创构思 The Scheme Design of Confucius Museum in Qufu 2006

城市与国土地方
Urban & Territory Places P107

P109　吐鲁番前进大队新农村住宅调查 The Loess Arch Residence in Turupan 1978

P111　智慧谷 原创桥 A Studio Design in the University of Manitoba -Wisdom Fountain & Creative Bridge 1990

P113　加拿大温尼伯中国城城市设计 The Urban Design for Winnipeg City Chinatown 1989-1992

P121　加拿大温哥华海滨地发展原创构思 The Concept for Vancouver City Waterfront Area Development 1992

P125　绍兴古越龙山城市中心广场原创构思 The Scheme Design of City Centre Plaza 1998-1999

P133　包头北梁晋风古城保护与发展 Baotou Jin-Style Historic Town Conservation & Development

P135　包头北梁晋风古城自然与建筑遗存调查 An Investigation on Baotou Town Natural & Historic Relics 2003

P151　包头北梁九江口核心区规划与城市设计 The Urban Design for Jiujiangkou Core Area 2003-2004

P171　九江口核心区14#地块规划与城市设计 The Urban Design for No.14 Plot 2004-2005

P187　包头东河水风景环境优化原创构思 A Concept to Upgrade East River's Waterscape Environment 2003-2005

P187　东河沿河土地城市发展环境策划意见 An Outline for East River's Waterfront Land Development 2004

P191　包头东河行政办公楼建筑方案设计 The Scheme Design of East River Administration Mansion in Baotou 2005

P195　传统晋风民居四合院优化发展原创构思 The Concept to Upgrade Traditional Jin-Style Siheyuan House 2004

P197　北梁两百年九江口财神庙城市广场重建设计 The Design of Jiujiangkou Caishenmiao Plaza in Baotou 2005

目录　contents

风景游憩地方
Landscape Sightseeing Places

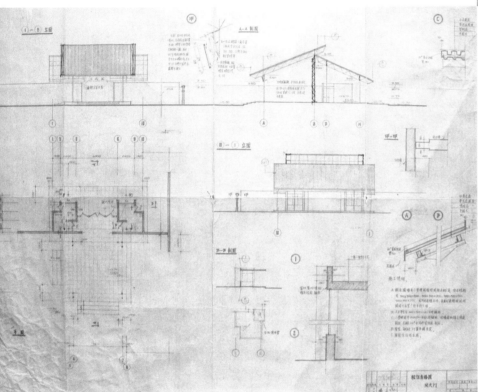

松江方塔园北大门 + 餐厅建筑与风景设计 1981-1985

The North Entrance Area Design of Pagoda Park in Shanghai: a continuation from tradition to the frontier of civilization relics garden design, directed by Professor Feng Jizhong.

松江方塔园由著名建筑学家冯纪忠老师主持原创,以一片建筑瓦砾中的"北宋兴圣教寺方塔、宋桥、明代城隍庙浅浮雕青砖照壁与河汊林地遗存为底蕴",将一些亟待保护的古代建筑迁入园中,打造以方塔为主体的现代古迹遗址园林。大门、钢柱与桁架结构支撑南北两片腾空斜面青瓦屋顶,覆盖下方公园入口青石围墙、建筑与青石铺地。北大门至塔院中心广场,古银杏绿地平台两段青石步行道、休闲平台与风景绿地,以几何曲线与直角锯齿交错契合的现代构图,东北角隅局部内层茶室,居反衬保护方塔、古建筑与林泉石的悠悠文史风情。庭院、回廊与厅堂峰回路转,砌石堑道与古银杏绿地平台交接处,大玻璃墙及青石地面构成纯、反差与融合。结构青瓦屋面,与白墙、青石墙、钢柱与桁架

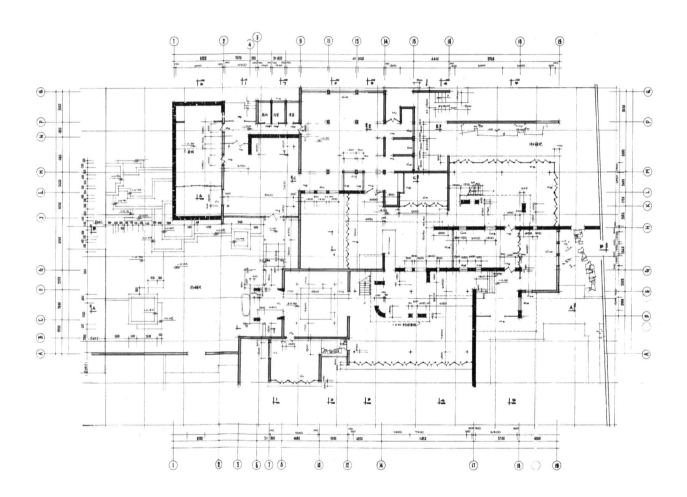

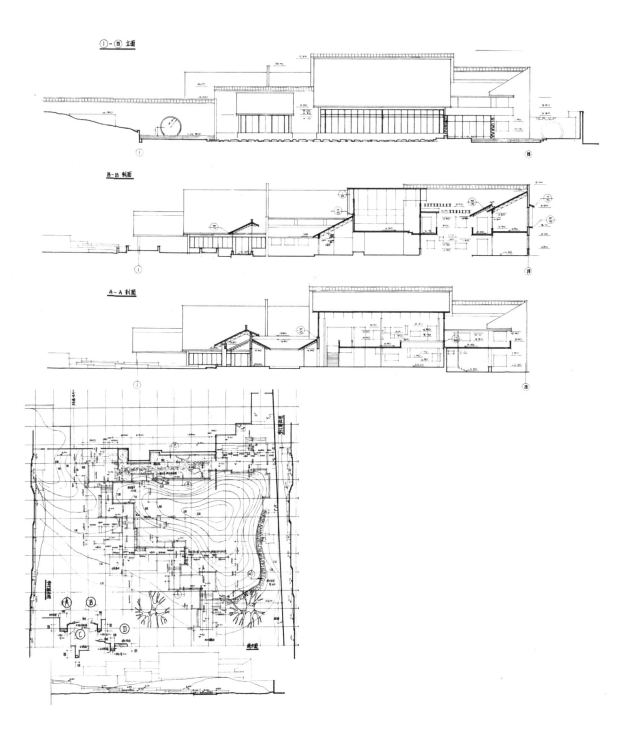

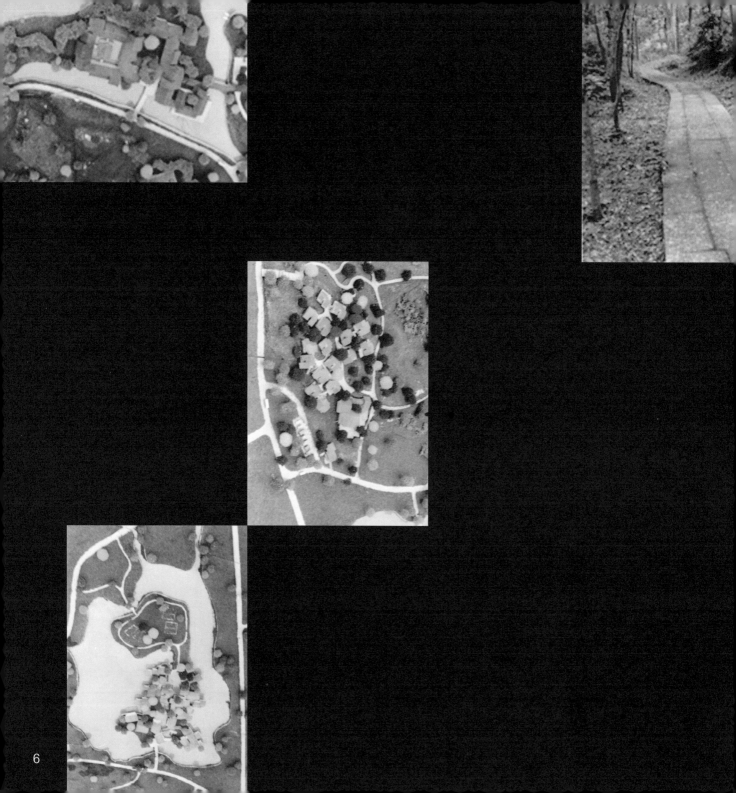

WEST LAKE HOTEL MASTER PLAN

杭州西湖国宾馆总体规划 1985

The Master Plan of West Lake National Guesthouse in Hangzhou: water, bamboo & villa, directed by Professor Dai Fudong, Chen Congzhou & Lu Jiwei.

1985年暑假至1986年，同济大学建筑学院戴复东、陈从周、卢济威三位老师带领，以"水""竹""居"为纲，进行向公众开放的临湖宾馆接待区与后山丁家山风景地方发展规划原创。临湖建"水庭院"和"水竹居"客房区，山麓沼泽地清淤泥、筑小岛、净水面，筑"湖上人家"客房区，山地辟"红叶山居"客房区，以修复古迹（康庄、摩星、礁石、鸣琴等）为中心发展丁家山风景地方。

大连金石滩风景区总体规划 1985

The Master Plan of Jinshitan Bay Scenery Area in Dalian: sea, island, green land, land form, sightseeing, sports, scientific discovery, entertainment & vacation.

同济大学建筑学院主持金石滩第一次总体规划，基于海、岛、沙滩、起伏草坡、植被与岩石地貌，策划建筑与风景地方的发展，开拓风景游赏、沙滩游泳、划船、垂钓、体育运动、科学探索与普及、娱乐与度假各种文化与旅游事业。

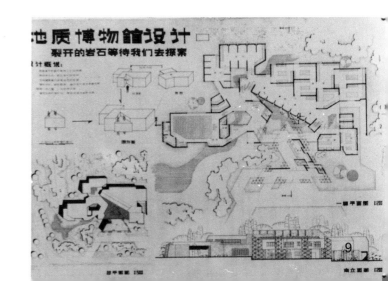

VALLEY SIGHTSEEING PLAN

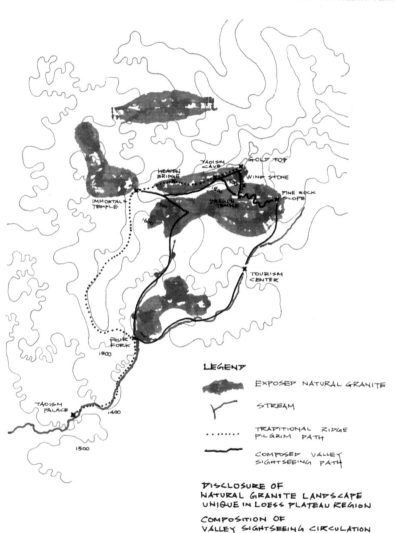

LEGEND

- Exposed Natural Granite
- Stream
- .... Traditional Ridge Pilgrim Path
- ———— Composed Valley Sightseeing Path

DISCLOSURE OF
NATURAL GRANITE LANDSCAPE
UNIQUE IN LOESS PLATEAU REGION

COMPOSITION OF
VALLEY SIGHTSEEING CIRCULATION

NORTH WUDANG MOUNTAIN

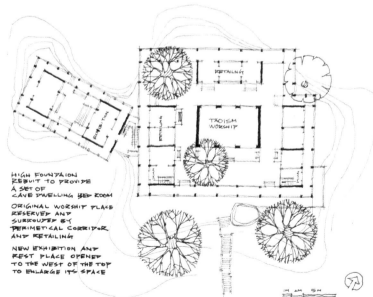

山西北武当山风景区总体规划 1986

The Master Plan of North Wudang Mountain Scenery Area in Shanxi: ancient path to Taoist Shrine, contemporary sightseeing line between rock walls, trees & rock fields.

传统游览线是道教朝山进香路：五里沙、五里沟、五里石级，直至金顶，而山顶寺庙建筑荡然无存。同济大学建筑学院主持第一次总体规划，从考察风景资源入手，发现了27km²与植被交错分布的天然花岗岩石风景旅游环境。修复金顶道教建筑，增加接待设施，在天然花岗岩石壁与林地交接部位开辟新游览线路，与传统朝山进香路线构成不重复往返的风景游览环路。

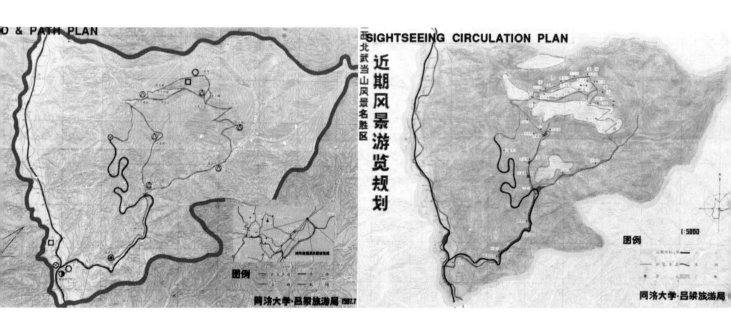

TERRACE AREA
CIRCULATION & TOURISM FACILITY
SITE PLAN

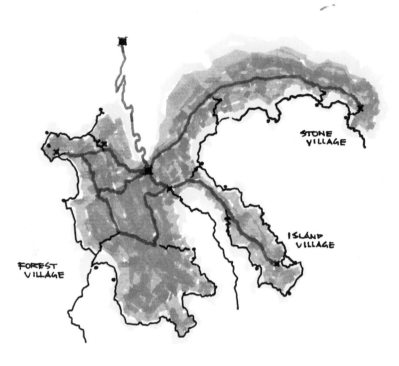

STONE VILLAGE

ISLAND VILLAGE

FOREST VILLAGE

LEGEND

- SANDSTONE TERRACE
- LIMESTONE TERRACE
- TRANSPORTATION ROAD
- SIGHTSEEING ROAD
- TOURISM CENTER

TOURISM FACILITY & CIRCULATION CONCEPTUAL DRAWING
TERRACE AREA
WULINGYUN NATIONAL PARK

武陵源国家公园总体规划 1987

The Master Plan of Wulingyuan National Park: master plan for first national park in China, grand limestone pit & sandstone peaks.

同济大学建筑学院李铮生、丁文魁、周秀堂、孔少凯四位老师指导武陵源国家公园第一次风景游览开发总体规划，覆盖张家界、索溪峪、天子山三个分区，1987年暑假期间完成14张风景游览与开发彩色手绘总体规划图纸与说明书及三个分区详细规划设计与说明书。

天子山为高原山地地貌，800m海拔以上为二迭纪石灰岩，800m海拔以下石家檐、神堂湾等地泥盆纪石英砂岩一度曾沉浸在海洋之中，均呈水平页岩分布。亿万年来，后者石英砂岩层逐渐被水与二氧化碳腐蚀，并遗留了数百座石英砂岩孤峰浮沉在山林云雾之中。最佳风景游览环境集中分布在天坑周边游览环线上，800m海拔以上的石灰岩台地，则以山脊游览线、石灰岩洞加垂直观景电梯，构成不重复的游览环路。

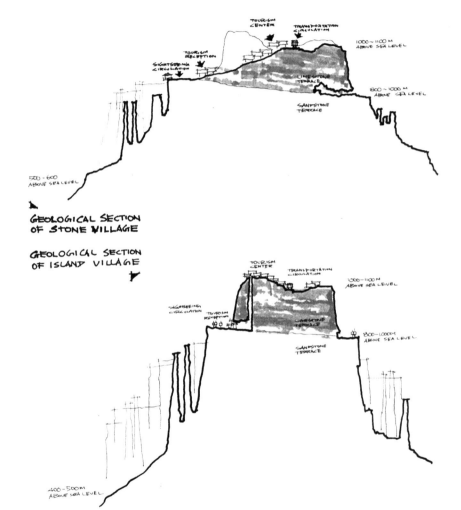

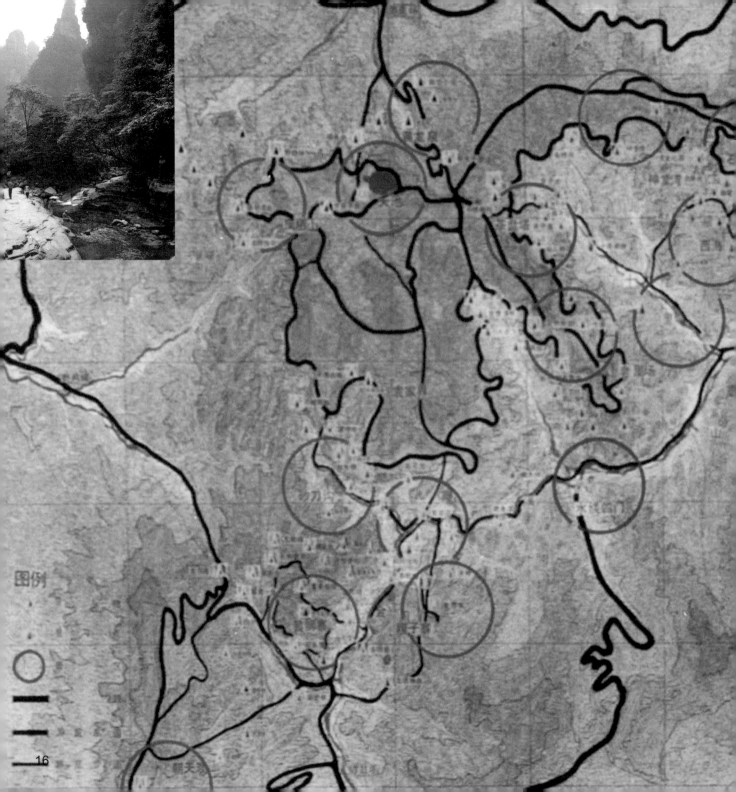

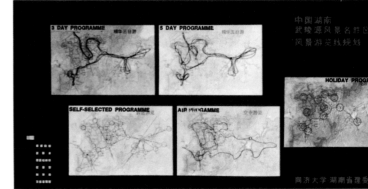

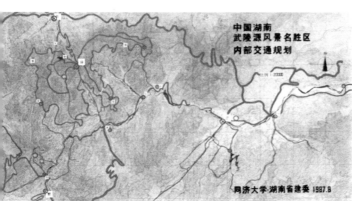

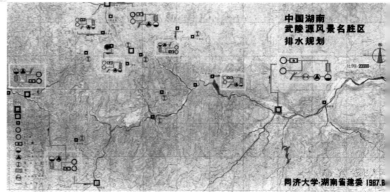

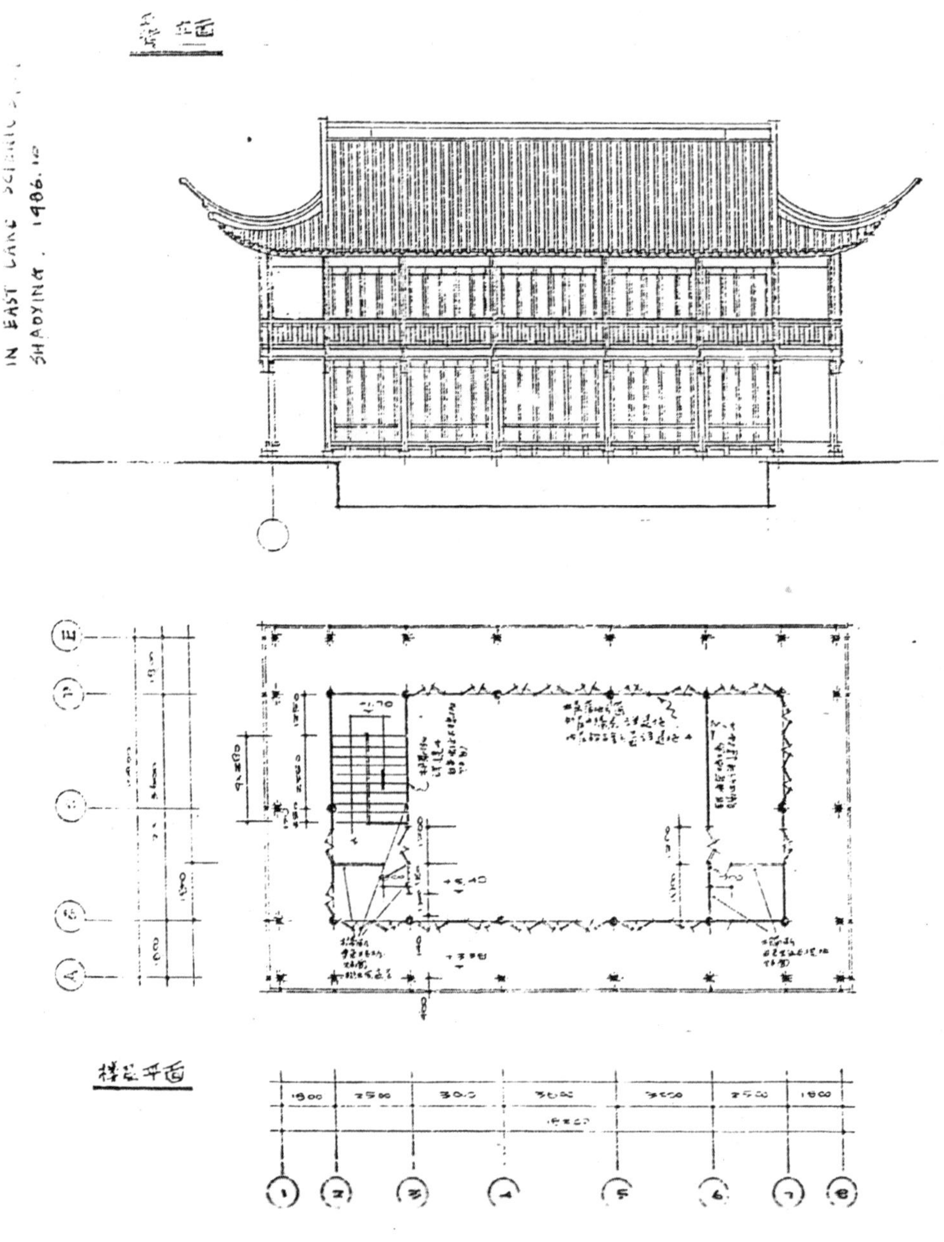

背立面

IN EAST LAKE SCENIC AREA
SHAOYING. 1986.10

楼层平面

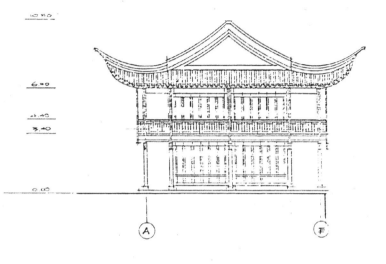

侧立面

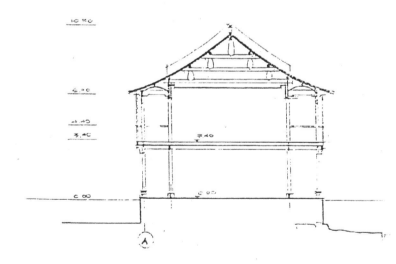

横剖面

绍兴东湖石宕稷寿楼古建筑设计 1985
The Pavilion Design in East Lake in Shaoxing:Traditional-style pavilion in a 2500 year quarry remains, directed by Professor Chen Congzhou.

东湖石宕是始于春秋越国的采石场遗址，拥有800m长近百米高采石遗存的绝壁、岩谷与沿绝壁一线的湖泊型深潭，是中华人借精心采石开拓山水风景名胜的大手笔。稷寿楼休闲茶室建筑设计是按古法营造法式在陈从周老师指点下做出的。

无锡太湖大浮山度假村总体规划与建筑设计 1997

The Design of Dafu Hill Resort in Taihu Lake: waterfront hillside site, a meeting place surrounded by sculptural forms of guestrooms, lobby & conference center.

一张8m×8m方格网，将会议中心、餐饮娱乐、跌落式客房区、总统别墅与接待中心建筑，全部定位在从北东南三面围合的一片濒临太湖缓坡迭落面西芳草坡地。系统发展理念加上锲而不舍的思考，使浸染诗意的动态三维数学模型与诗意生存环境水到渠成：跌落式客房群、主入口与总服务台、会议中心与餐饮娱乐三块，从西、东北与东南三面围合一个正圆形平面室外柱廊平台与平台下方的多功能宴会厅，构成建筑与风景、聚会与交流的核心。

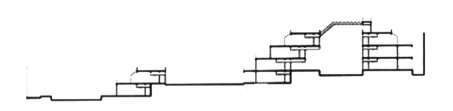

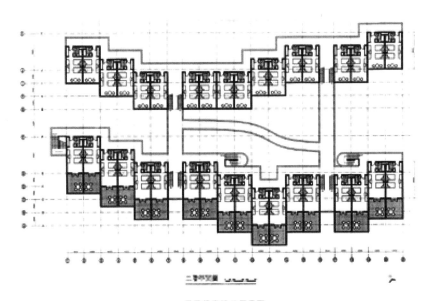

四星级宾馆二层平面

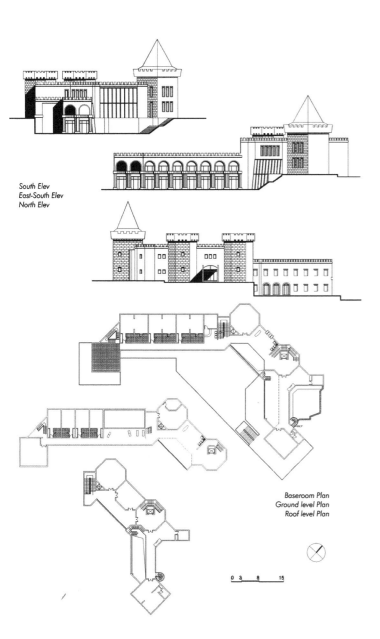

South Elev
East-South Elev
North Elev

Baseroom Plan
Ground level Plan
Roof level Plan

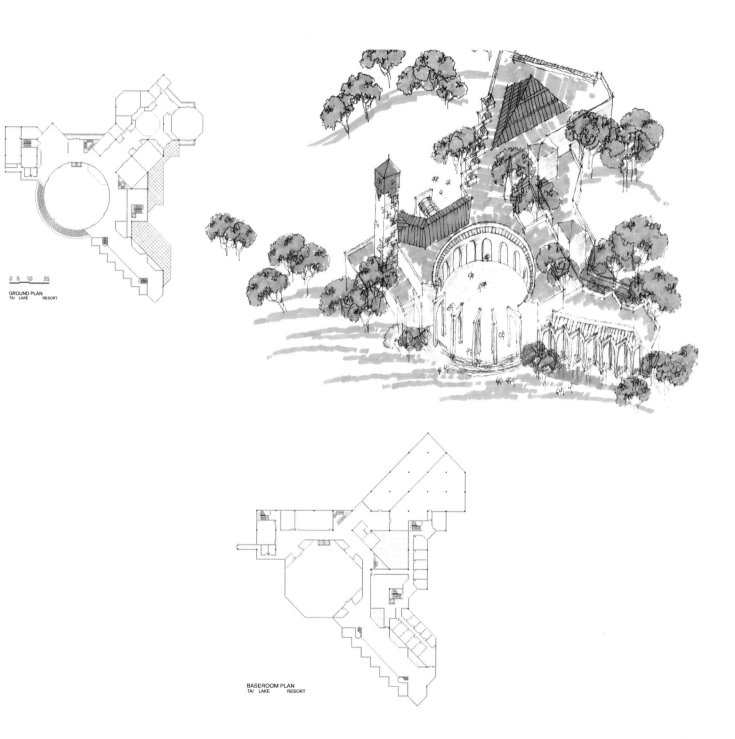

GROUND PLAN
TAI LAKE RESORT

BASEROOM PLAN
TAI LAKE RESORT

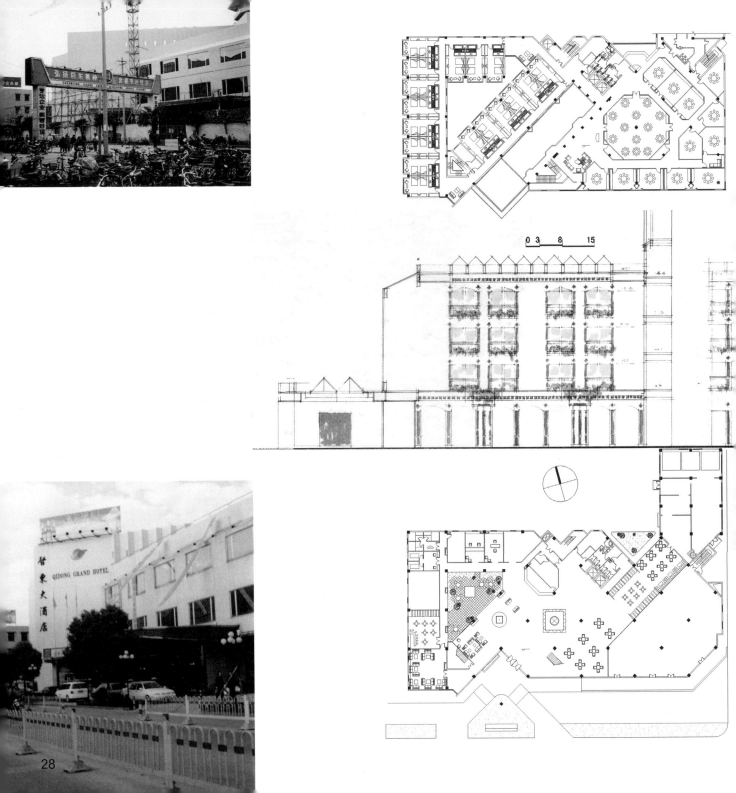

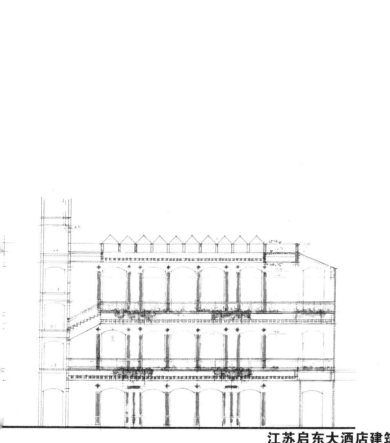

江苏启东大酒店建筑设计 1998–1999

The Design of Qidong Hotel in Jiangsu: Whole experience-scape, atmosphere, environment, transformed by an off-center main axis.

一次冲决直角坐标系动态三维图式思维的探索：建筑南北中轴线向西南偏转45°，一着棋变全局动；与马路成45°角抽象古典矩形平面三层高室内中庭大堂，西侧单廊客房围绕一等腰直角三角形平面玻璃天窗室内中庭布置，东侧餐饮娱乐区涌现偏转45°角的包厢、走廊与正八边形宴会厅。

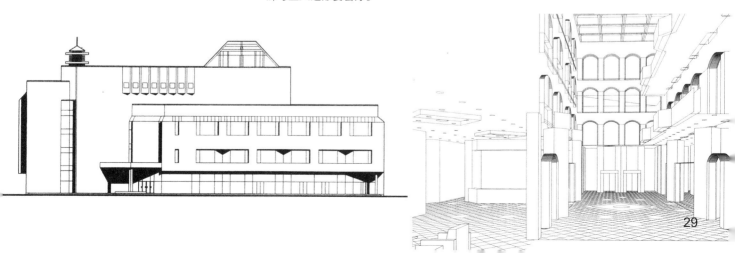

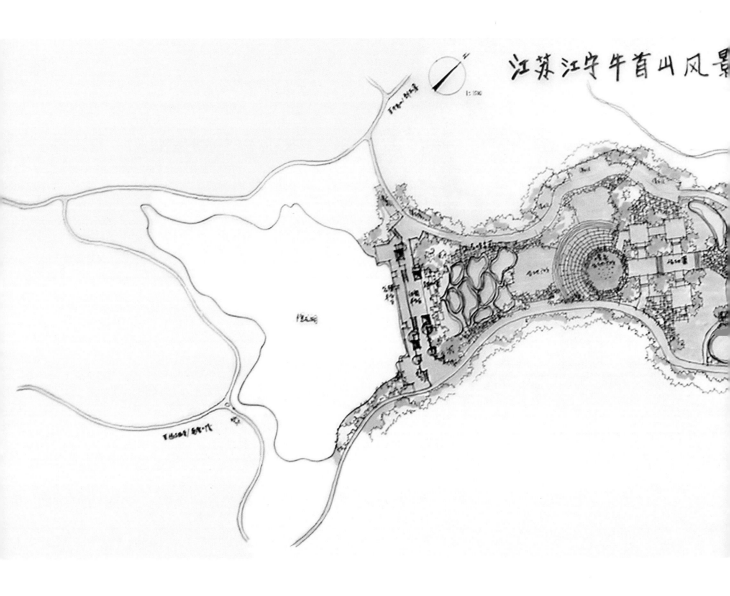

江苏江宁牛首山风景

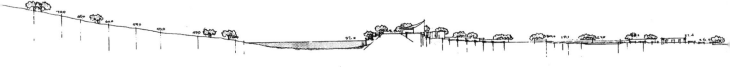

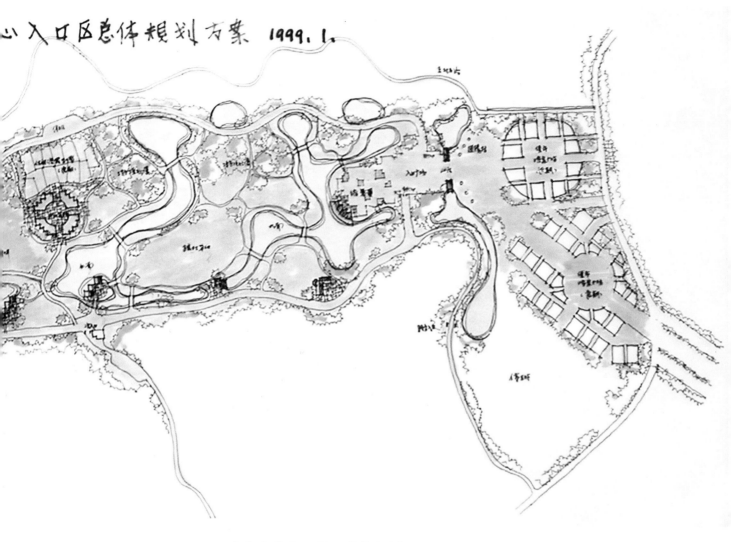

南京牛首山风景区总体规划 2000
The Master Plan of Niushou Mountain Scenery Area in Nanjing: 10km² reservoir lowland water soil greenland plan.

10km²基地开发原创构思，含牛首山水库以及其下游溪谷山林水土地。利用大坝高程落差打造岩溶地貌、台地园与露天剧场，馆舍散点曲溪平湖山林之间，酿成旅游、文化与餐饮娱乐地方。

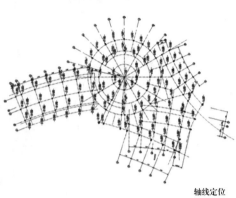

轴线定位

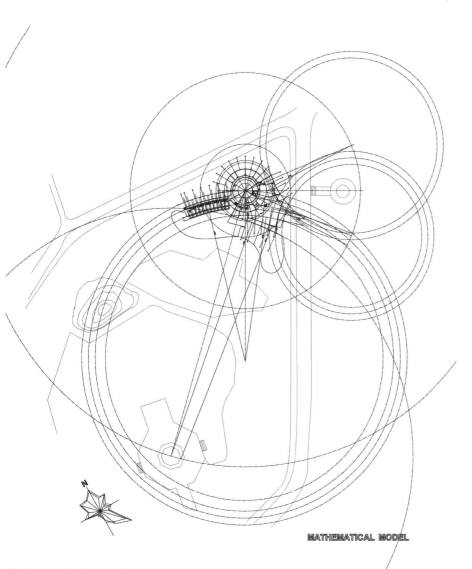

MATHEMATICAL MODEL

启东鼋头角海滨度假村建筑设计 2002-2003

The Design of Yuantou Cape Seashore Club: a 60° clipping angle water soil site enclosed by two sea dykes, irregular built form of circle, arc & free curves.

建筑总体几何模型是基地风景构成理性分析的结晶：两条海堤围合60°夹角水土地，围绕圆形平面水庭院的圆环形平面接待大堂应运而生，是个人建筑总体几何模型原创思维的最新发展。跌落式客房带南向弧形平面与餐饮娱乐部西南、东南、东北向弧形平面，如风行水上、自然成文。圆、弧与自由曲线成为形态塑造、空间构成与风景网络原创的主题，规则几何形演变为自然几何形。

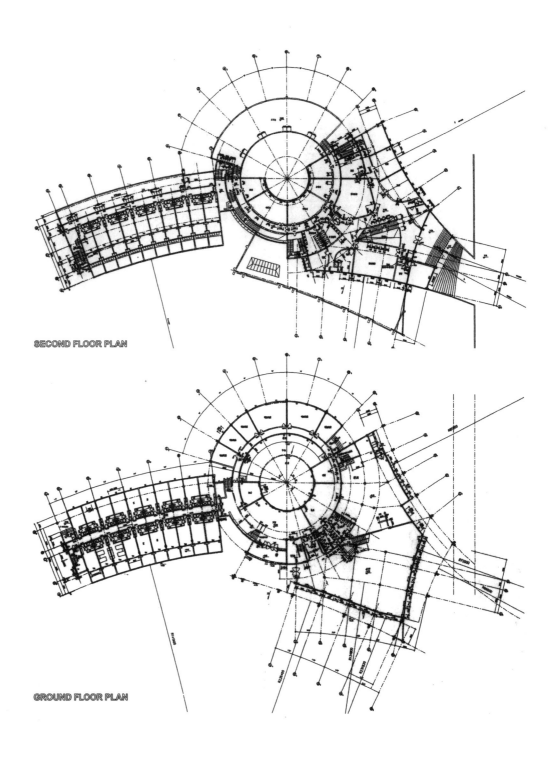

LOBBY/COURTYARD SECTION

LEISURE AREA SECTION

GUEST AREA STAIR SECTION GUEST AREA SECTION

EASTNORTH ELEVATION EASTSOUTH ENTRANCE ELEVATION

SOUTH ELEVATION

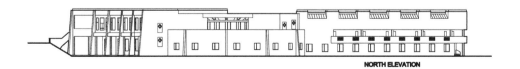
NORTH ELEVATION

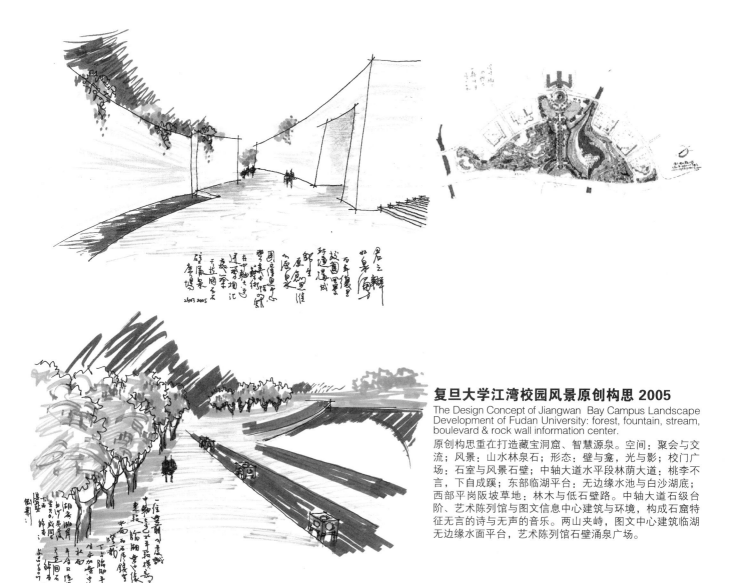

复旦大学江湾校园风景原创构思 2005

The Design Concept of Jiangwan Bay Campus Landscape Development of Fudan University: forest, fountain, stream, boulevard & rock wall information center.

原创构思重在打造藏宝洞窟、智慧源泉。空间：聚会与交流；风景：山水林泉石；形态：壁与龛，光与影；校门广场：石室与风景石壁；中轴大道水平段林荫大道：桃李不言，下自成蹊；东部临湖平台：无边缘水池与白沙湖底；西部平岗阪坡草地：林木与低石壁路。中轴大道石级台阶、艺术陈列馆与图文信息中心建筑与环境，构成石窟特征无言的诗与无声的音乐。两山夹峙，图文中心建筑临湖无边缘水面平台，艺术陈列馆石壁涌泉广场。

公共建筑地方
Public Building Places

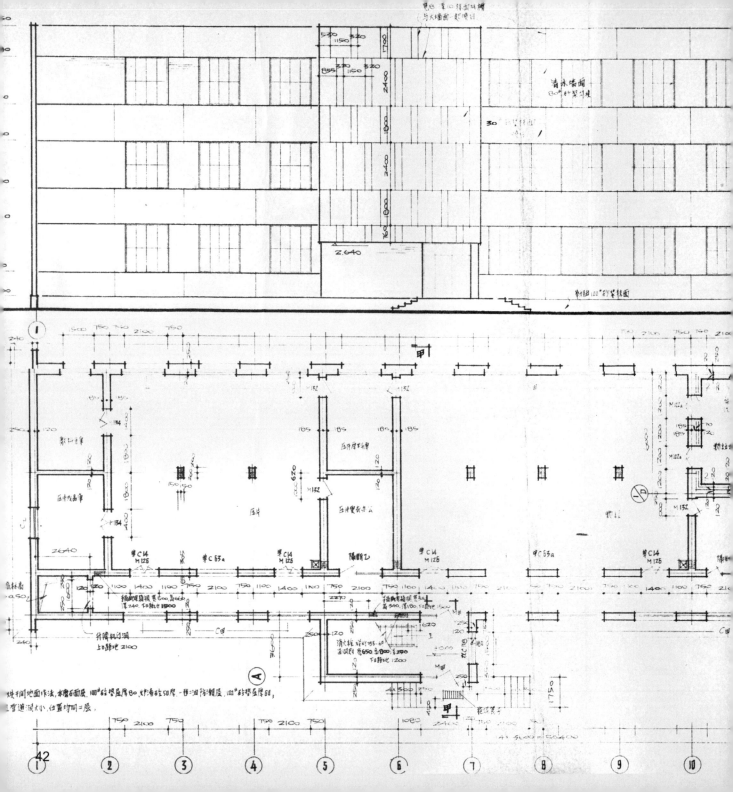

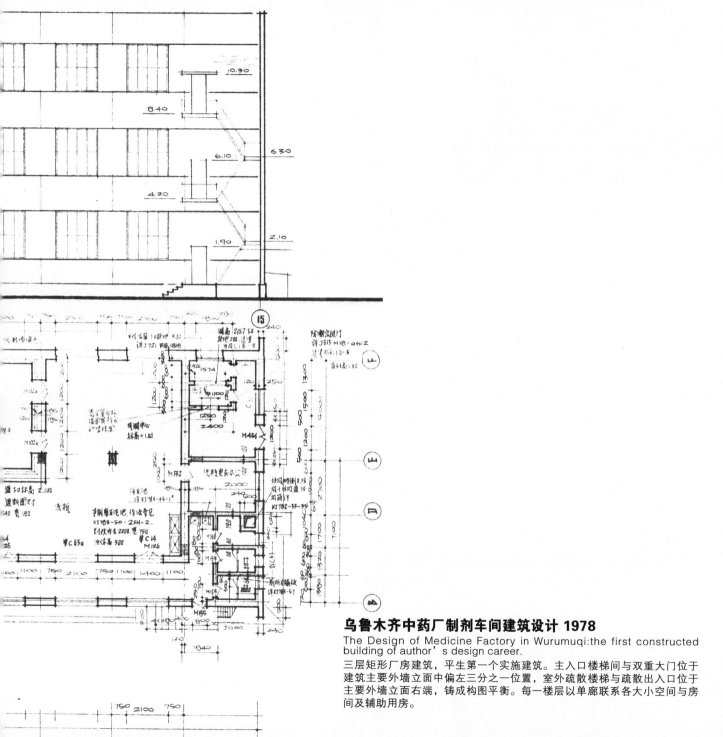

乌鲁木齐中药厂制剂车间建筑设计 1978

The Design of Medicine Factory in Wurumuqi: the first constructed building of author's design career.

三层矩形厂房建筑,平生第一个实施建筑。主入口楼梯间与双重大门位于建筑主要外墙立面中偏左三分之一位置,室外疏散楼梯与疏散出入口位于主要外墙立面右端,铸成构图平衡。每一楼层以单廊联系各大小空间与房间及辅助用房。

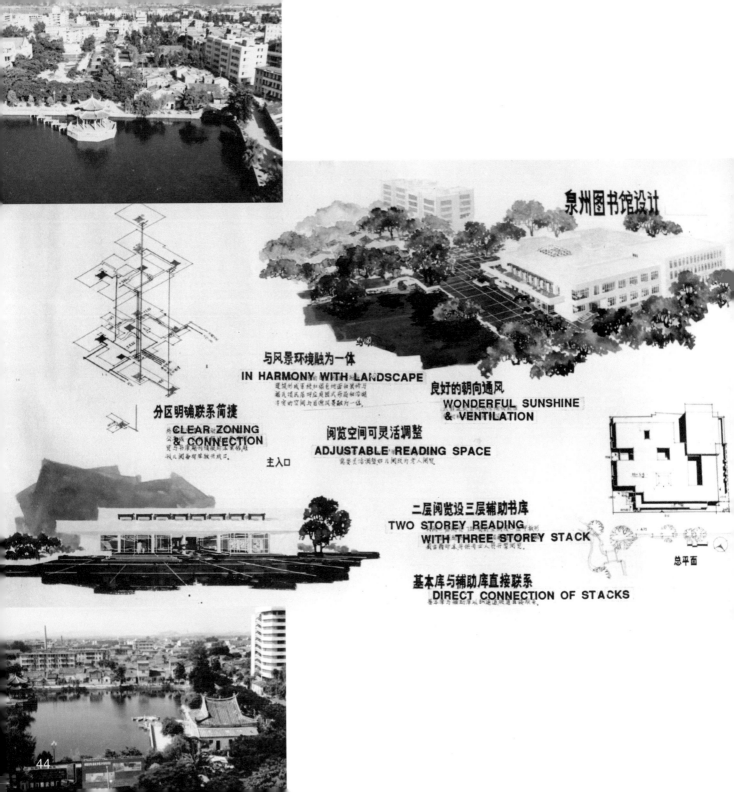

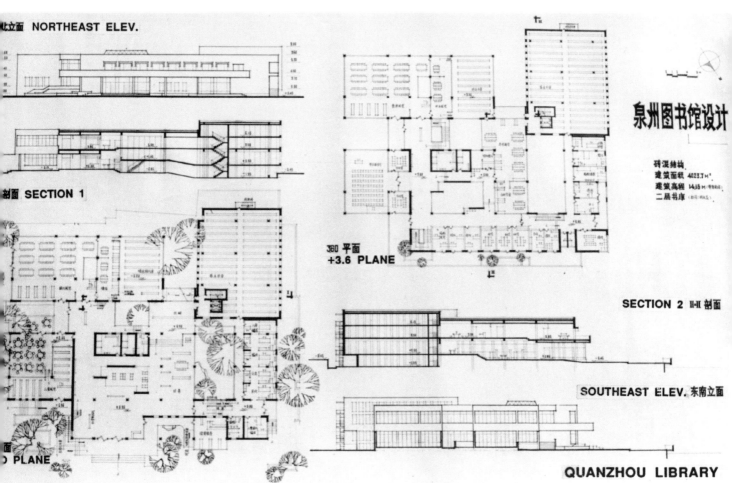

泉州图书馆全国获奖建筑设计 1983
An Award-winning Scheme Design for Quanzhou City Library National Design Competition: integrating library, temple & lake.

全国邀请赛三等奖获奖方案（无一等奖）。庭院式建筑与百源清池古寺庙及风景水面融为一体，取得良好的采光与通风。书库、阅览与交流陈列地方，通过庭院与室外回廊进行分隔、联系与调节。

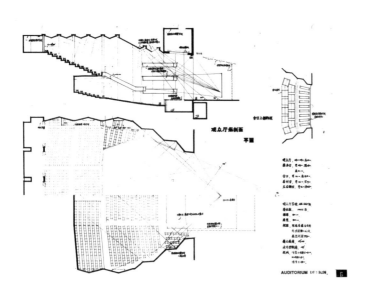
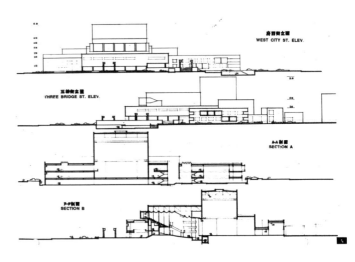
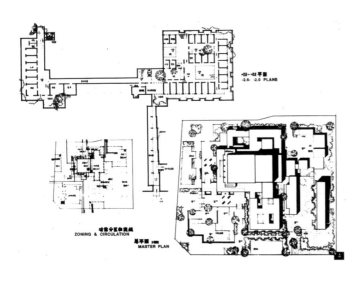

现代建筑功能和造型

质朴高低错落大块笨体组合的造型突出了现代剧场建筑的观演功能,休息厅二层择利山西风景名胜从突墙面高侧窗接收有利环境和传统建筑特征并成不同照度环境之间的过渡空间休息后专层平台形成共赏空间

传统建筑布局特征

等用中国传统山野建筑(如恒山悬空寺五台山佛光寺等)因地制宜的布局特征,主体建筑居于中轴线上,而入口空间偏离中轴线并采用门栏作为入口空间标志。

石窟环境特征

墙红色下沉式外部空间单个平台提供从不同标高和角度观察剧场错落造型的空间平台石壁作摩崖表现出山

传统文化开辟小吃方饮小商环境,与天龙山空底石窟和黄土高原下沉式庭院相呼应。

优越的视听环境

阶梯座后6排外某余138座全部处于22m视野之内池座1一9排C=6cm,14-19排C=9cm,座后C=13cm视众后台上方使大堤反射面增强前中前座池水剧声,观众厅前后设恩精调节空间适应多功能需要有效改善光最少自然半能摄取打浦东池台提供博出式墓台之廊一道面光从适应表演区位置变化。

合理的分区和流线

分区明确联系便捷观众迅进场疏散流线路通畅场时错场与排散后演出题回量贵环秋客台工作流线平行分装互不交叉与观众后后台工作位演出准备与剧场水

公挟者方便满工往回空间避台二侧有段设并地下篮球培道路。

演员心理和活动空间

根据演以创作的整体性和演员思想能通通高度请中的联业者典设计上下日展贵避的演出任密设施装置通场体训空间分层台层楼层本地下五层。

演员雕塑的事业心剧务的竖习请救泽出的堂隧俊之以大量时间和精力在某生活空间中从事地下的艺术创作活动演以从职场为主的生活空间提供爱身有用的条件流通请求声底某万个人腿功成果空间场遇信息分层平台体小组临月排排接好的生活空间宿塘演员某亚如目标成对剧场建筑得缓之一。

☆ 国有总体环境设计占地范围设大请示活于杆网志转商发部行应度闲意平立,剖图按1:250比例绘制。

TAIYUAN THEATRE DESIGN
太原大剧院设计

太原大剧院全国邀请赛建筑设计方案 1984
A Scheme Design for Taiyuan City Theatre National Design Competition:temple, rock wall, sinking plaza & quarry.

建筑基址位于迎泽大街城市中心。基于演员与观众两个群体活动功能与审美体验,打造合理的流线与分区、优越的视听环境与出类拔萃的功能地方和现代(非古典)建筑造型。采用恒山悬空寺等山野建筑因地制宜总体布局特征:主体建筑位于总平面中轴线上,而主入口偏离主体建筑中轴线并以山门为标志。建筑群东南街角筑下沉式窑洞式商业文化广场,与穿越城市主干道的地下人行横道和剧场建筑群融为一体,共同构成高低错落富有雕塑感的石窟风景环境。

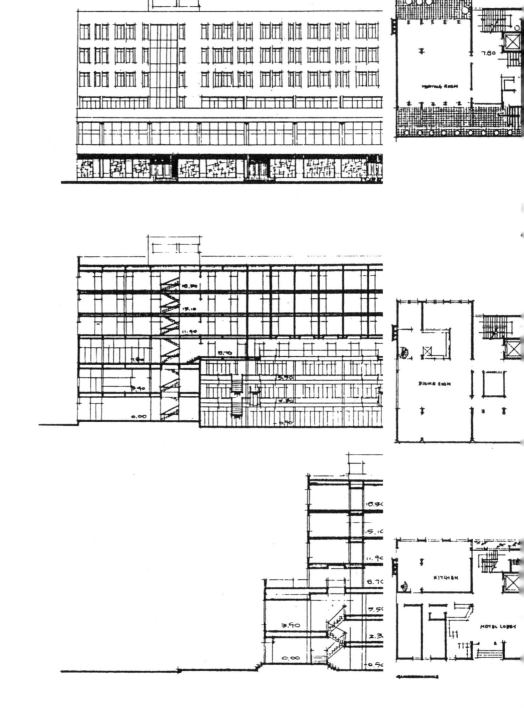

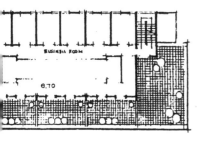

绍兴烟箩洞宾馆+商业展示中心 1985-1986
The Design of Yanluodong Hotel & Shopping Centre in Shaoxing:exhibition, interior courtyard, complex floors & free form entrance.

涵盖商业供销中心与六层接待旅馆。商业展销中心四层高室内中庭，南北错半层以步行楼梯联系，构成供销与陈列最佳氛围与环境。地面层两个中心落地长窗大门从外墙后退并偏转20°，辟斜向直角三角形平面门斗，构成迎接并容纳人流的最佳空间与风景。

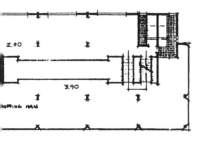

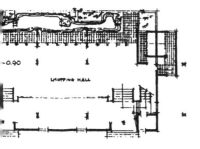

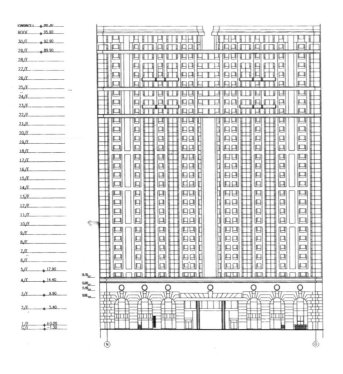

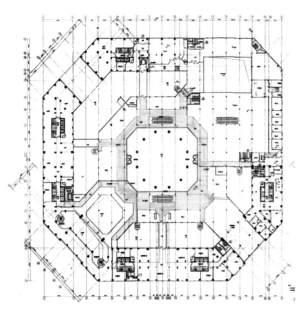
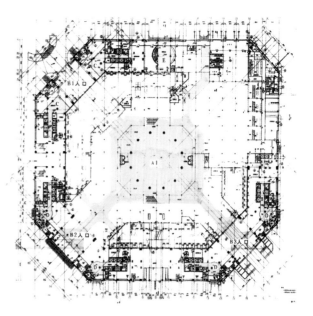

上海海友花园光大会展中心建筑群设计 1993–1997
The Design of Everbright Buildings Complex in Shanghai:200,000 m², 6 high-rise buildings & podium,architecture,landscape & exhibition centre design directed by Dr.Tao Ho.
20万m²酒店、商住、住宅、商业与会展中心建筑群，在正方形基地上东北、西北与西南45°转角布置三座凯旋门，六幢100m高层建筑，两层满铺地下车库，周边三层高裙楼围合32m直径球面玻璃天窗室内中庭。香港何弢建筑设计所主持建筑、外墙、室内与外部环境设计，与华东建筑设计院、机电四院上海分院合作进行施工图设计。

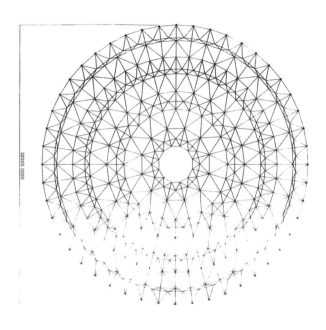
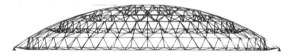

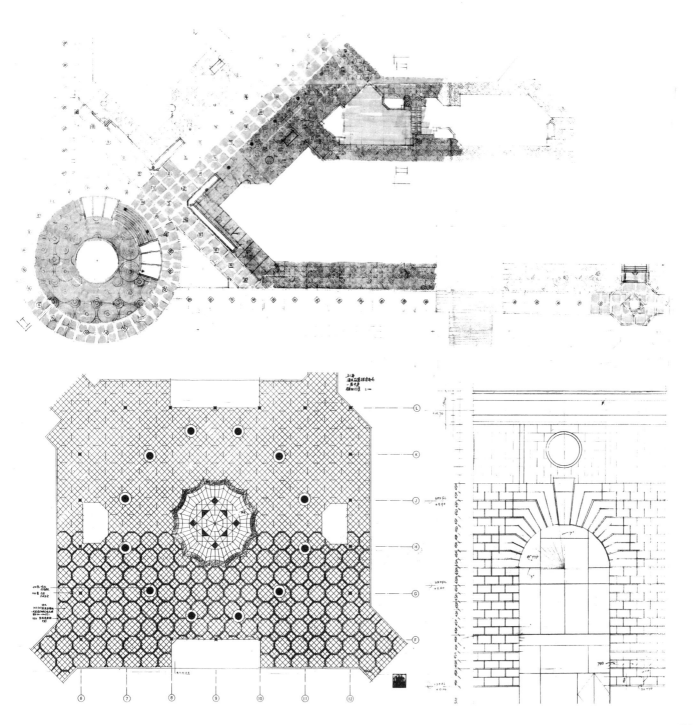

55

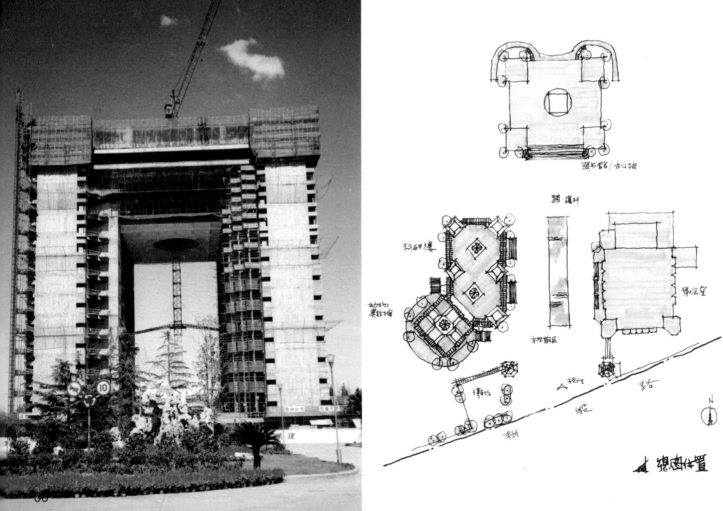

上海第二军医大学图书馆办公主楼+科学实验楼设计方案 1994–1997

The Design of Library & Science Buildings, No.2 Army Medicine University in Shanghai:Triomphe–style main mansion, directed by Dr. Tao Ho; 45° cutting angle pattern, reflecting a molecular structure.

翔鹰路校前区广场凯旋门式大学主楼，正方形平面，银灰色玻璃幕墙加天山红石材外墙，正南有4.8m高大台阶。西翼行政办公楼、东翼图书馆，11至15架空层为会议接待。两层高入口大台阶，锯齿形或转角通高落地玻璃幕墙长窗，与从屋顶层逐层后退落地玻璃幕墙长窗相映成趣，相得益彰。香港何弢建筑设计所主持建筑与外墙设计，上海纺织工业设计院合作进行扩初与施工图设计。

西安第四军医大学教学 + 图书馆主楼原创构思

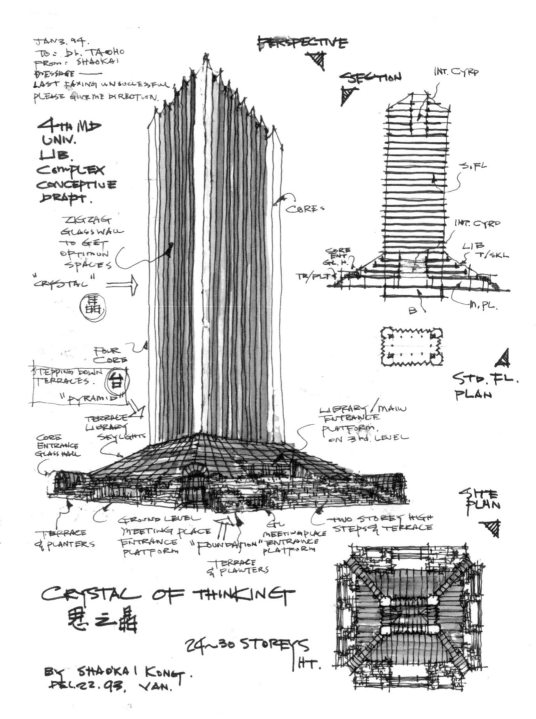

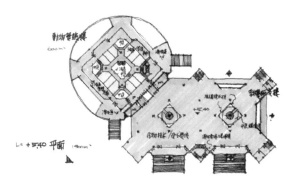

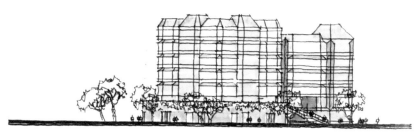

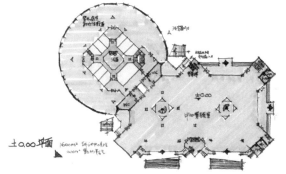

（接P57）与第二军医大学会堂中轴对称的科学实验楼位于广场西侧，平面以正八边形和45°切角为母题。建筑矩形平面南北向布置，宽阔的大台阶直上二层凹入口平台。结构采用8.4m×8.4m柱网体系，轴线按正南北偏转45°布置，建筑凸显多个45°外墙雕塑。

动物实验楼正方形平面、南北中轴线向西偏转45°布置，底层以圆形围墙圈定外部空间作为实验动物放养场地。引进天窗采光室内中庭与玻璃陈列橱窗，取得最佳的聚会地方、风景和楼面价值。外墙采用水平向花岗岩饰面外墙、落地长窗加45°转角斜向玻璃天窗。建筑造型表现为多个正八棱柱的组合，表现了医学和药学的分子结构境界。

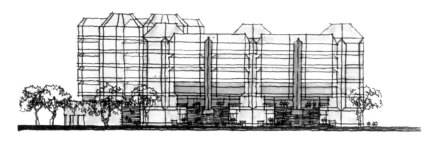

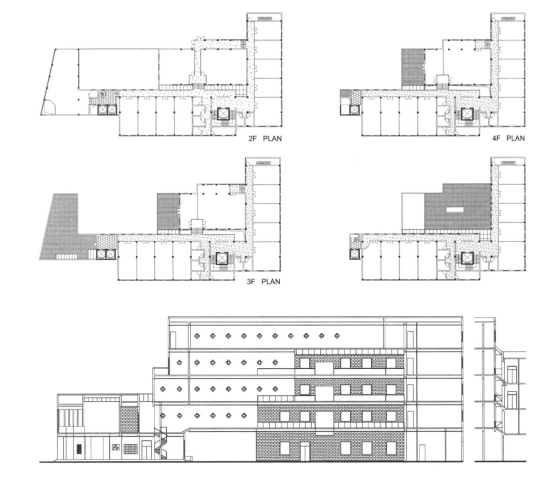

2F PLAN

4F PLAN

3F PLAN

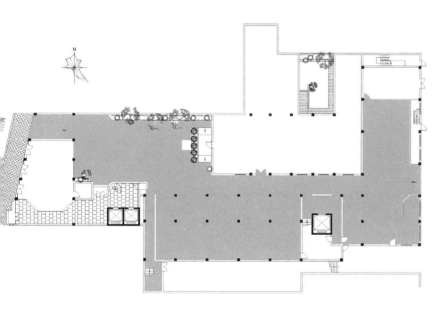

上海照相机四厂改建设计 1994-1995

The Renovation Design of No.4 Camera Factory Buildings in Shanghai:upgrading factory shops to high-end office.

香港何弢建筑设计所主持设计,将地处淮海中路黄金地段照相机四厂厂房打造成国际标准高档办公楼:东楼、南楼与北楼结构加固,重建沿襄阳南路的西楼,进行整体室内与外墙设计及各层楼面空间布置策划。新建西楼办公区入口门厅仅留下一条2.2m面宽的狭长过道,西端玄关和东头电梯前厅,分别筑偏转45°的黑色大理石陈列墙面和3.9m至4.2m面宽两层高天然采光中庭。改建工程的核心,是将北楼南楼东楼之间东西长23m、南北向间距仅1m、二至五层高的狭长室外天井与外廊,升华为泰山面砖墙面天然采光中庭,构成整个办公区聚会与陈列地方。其精华则是在东楼南楼交会处,发现并恢复3m进深天井的初建原貌:一条小溪在峭壁转折处汇成清潭,大大增加天然采光量。

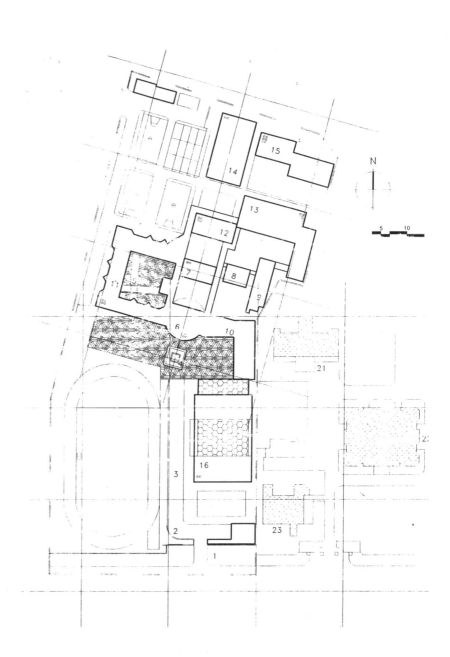

绍兴中学百年校园改建方案 1994

The Concept Design of Shaoxing Secondary School: overlap & deviation of form & style, re-architecture plan for author's alma-mater, started by Professor Cai Yuanpei.

1907年始建校园南临水澄巷、府河和仓桥，距古越龙山仅半华里。其中硬山顶两层五开间校长楼（蔡元培、鲁迅工作居住过的地方）、七开间女生宿舍楼和两层三开间歇山顶医务楼楼均为1907年建，仍保持粉墙青瓦深红色木结构，青砖铺地，庭院穿插其间的浙东书院风情。

百年改建原创构思，表现叠合与偏转的数学模型：基于西面试弄道路走向，校园总平面由南北两个互相偏转15°的坐标轴平面叠合而成。S平面操场体育馆建筑采用与东邻市府相近的现代风格，下自成蹊林荫大道，南端为百年树人纪念照壁。北部Z平面定位中部教学办公区与球场宿舍区，以校长楼为核心，传承浙东书院粉墙青瓦、长廊与庭院传统气韵。中心广场S、Z两坐标轴平面在教学区与操场区交接部交会。大面积铺地广场，门厅以东三层梯形平面建筑辟为梯形阶梯教室。中心广场5m×5m平面玻璃幕墙建筑，主入口门厅圆形金属纪念框架以Z平面作为构图参照系，而下方铺地纹样则以S平面为构图参照系。

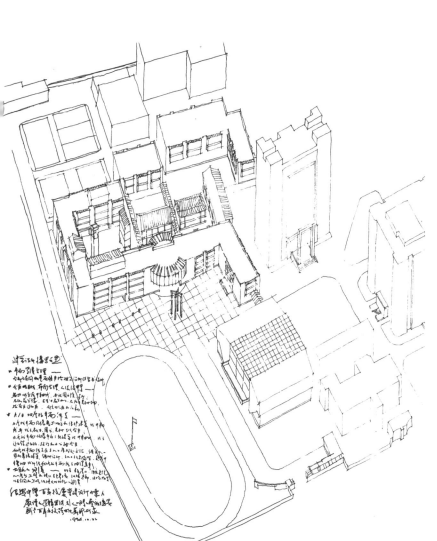

上海外滩光大银行改建设计 1998

The Renovation Design of Everbright Bank in Shanghai Bund: refreshing the interior of Banque De L'indo Chine Building, new building's elevation abstraction.

面向外滩主楼原为法国东方汇理银行，1907年按巴黎小特里亚农宫式样建造：外墙为爱奥尼克柱式三段式立面，底层为营业大厅，中央扁平玻璃天窗，二、三层办公生活用房环绕中央天井和内走廊布置。

主楼改建的核心是保护1907年建筑遗存：底层为营业大厅，二楼辟国际金融业务区，三楼天井加建钢筋混凝土楼板、筑浅拱形玻璃天窗，作为行长办公室。底层大厅取消加建夹层，恢复爱奥尼克柱式墙、柱、檐口，银线米黄大理石饰面。拱形木窗改为古铜色铜窗，保留并出新1907年遗存木雕山花大门套门、樱桃木门窗套厅和木雕壁炉。大厅中央建浅拱形方格人工采光玻璃天棚，四周白色天棚抽象古典构图纹样，源自巴黎罗浮宫及西欧采风手稿，并融入中国古典风格。三楼中央接待厅及周边办公和会议室，采用与门套同高的古典樱桃木墙裙。

新建辅楼南、北外墙立面，是对主楼爱奥尼克三段式立面构图的抽象。辅楼南北宽11m、东西长34m，含半地下层车库和地上五层办公服务区。建筑平面东西向依次剪裁为门厅、公共走廊、电梯、开敞式办公区和附属用房。辅楼形体剪裁最重要的一笔是，基于对交通流线分析，在主辅楼接合部辟出矩形平面辅楼主入口会聚空间：公共走廊南侧蓝绿灰色玻璃幕墙向北后退3m，结构框架暴露，构成雕塑+光影的辅楼主入口外部空间。

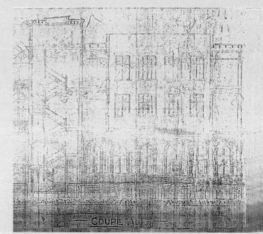
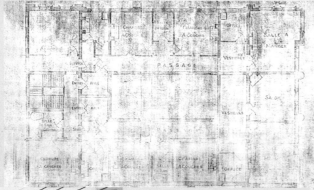

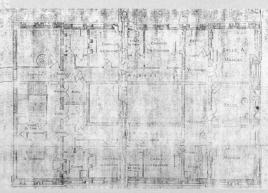

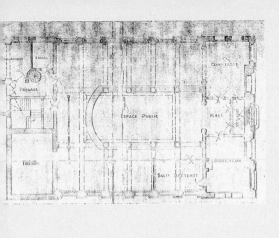

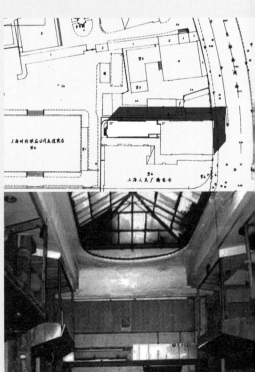
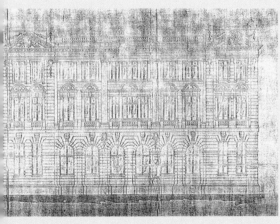

HALL RENOVATION PROPOSAL
EVERBRIGHT BANK RENOVATION

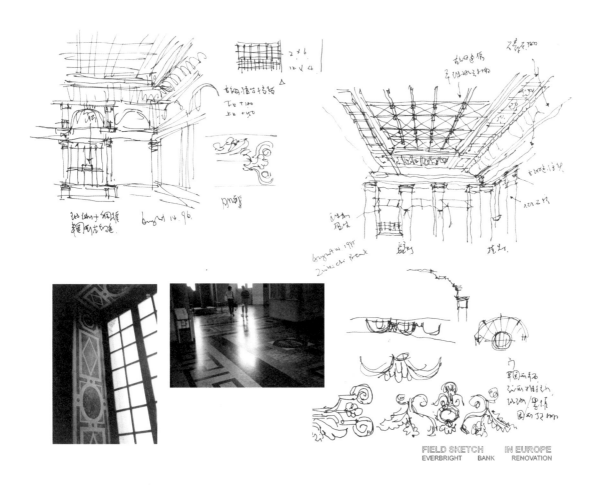

FIELD SKETCH IN EUROPE
EVERBRIGHT BANK RENOVATION

Longitudinal Section

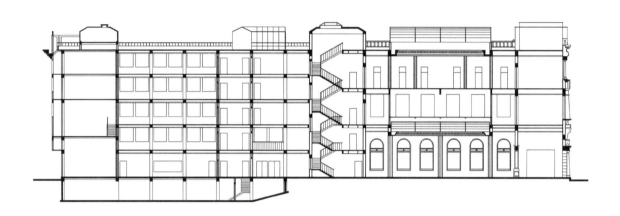

Ground Level Plan

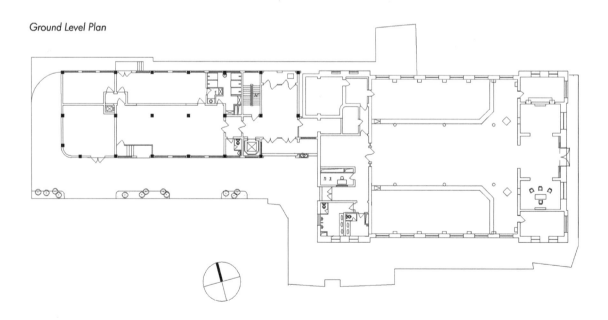

Suzhou River Elev

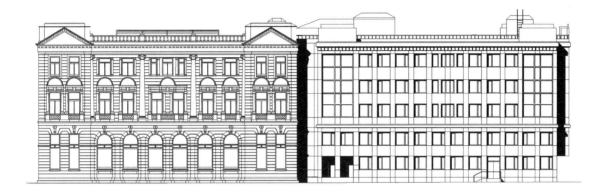

South Elev

Roof level Plan

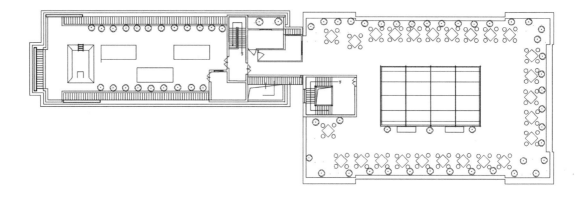

3F(old)/4F(new) Plan

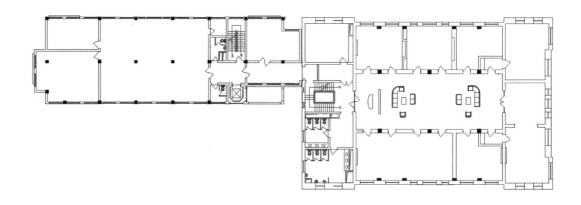

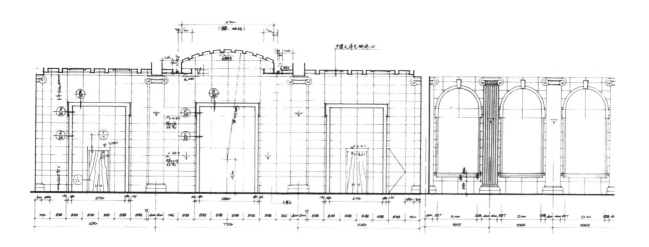
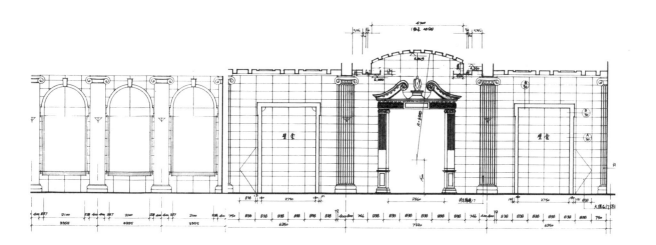

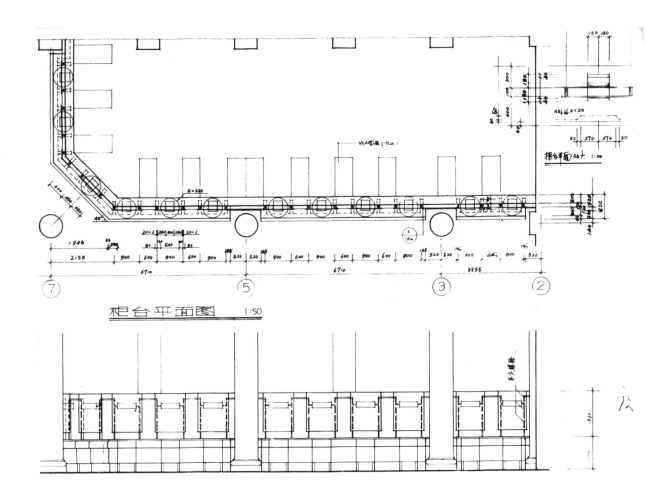

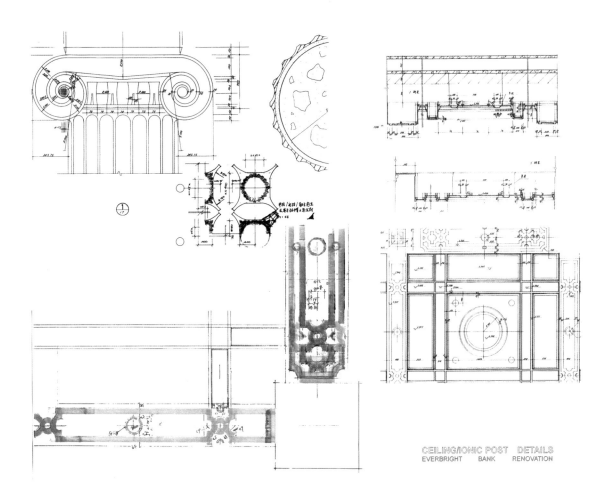

CEILING/IONIC POST DETAILS
EVERBRIGHT BANK RENOVATION

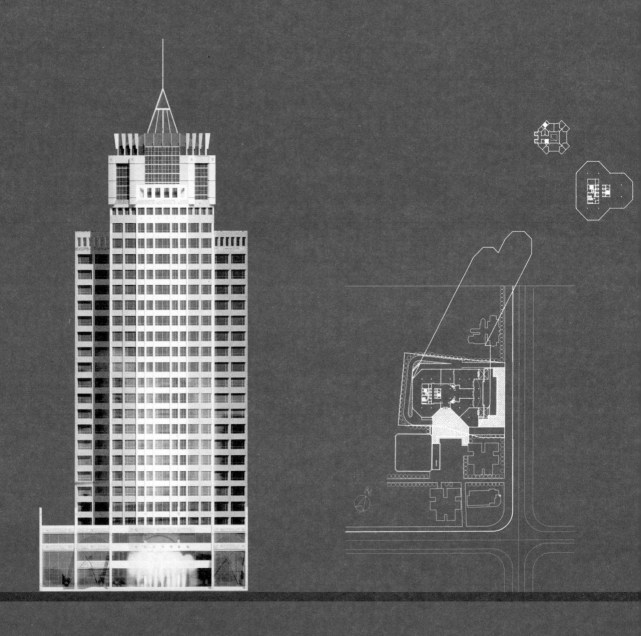

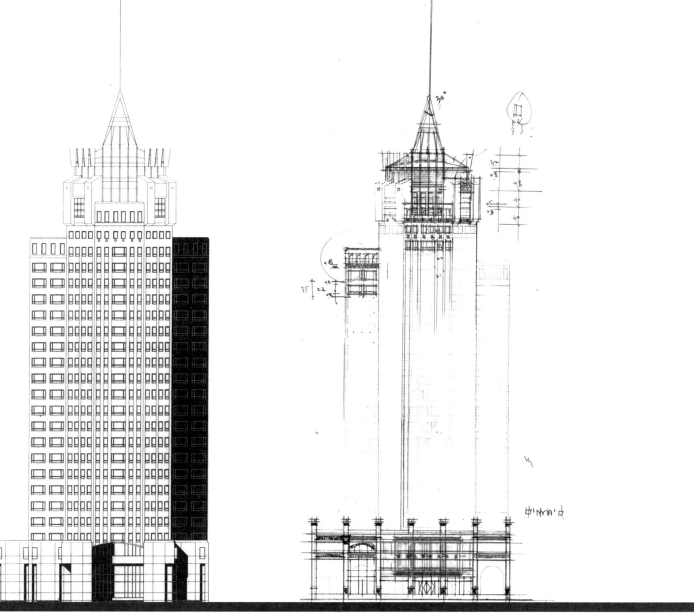

上海爱建大厦建筑设计方案 1998

The Scheme Design of Aijian Mansion in Shanghai: street-front office-units, sculptural building form & roof-top.

方案原创重在资产价值评估与建筑造型。1至3层为宛平南路门厅采用天然采光,室内中庭为商业与交流地方。高层办公楼建筑平面为两个八边形相套:后部4至23层扁宽八边形构成前部正八边形的背景与底蕴,45°大切角平面增加高价值临窗办公单元。24至27层总裁办公与会议区,是建筑顶部造型的精心雕塑。24至26层楼面45°大切角外墙实为主虚实相间;24层四片外墙向中心退缩,构成拥有通高落地门洞系列的屋顶花园。27层正八边形平面玻璃幕墙到屋顶檐口转为斜面玻璃天窗;在幕墙与天窗檐口转折一线做4m高不锈钢微外挑梳子形雕塑,建筑最高屋顶平台构筑不锈钢向天尖锥。

上海大学美术学院建筑原创构思溯源影像 P84左上2500年龙游采石宕遗址 左下巴黎奥赛博物馆室内
右下8000年阴山岩画 右上香港中文大学建筑教室后墙采光落地长窗

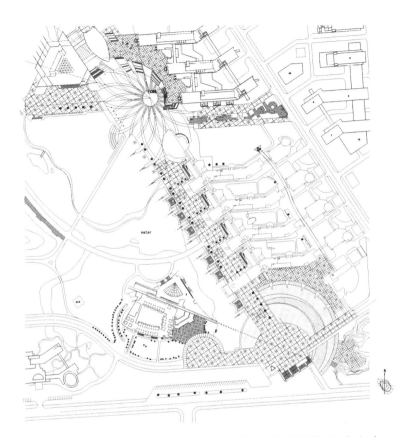

上海大学美术学院建筑设计方案 1998

The Scheme Design of Art Institute, Shanghai University: grotto space & rock sculpture, meeting & exchange, wall & niche, interior court-yard, corridor fountain & outdoor court-yard.

原创构思源自2500年采石场遗址逐层开采水平层理石壁与洞窟：石窟环境，巨石造型。

建筑总平面造型雕塑是周边建筑和视线交汇融合的结果。沿着从校前区广场中心点看美术学院的视线筑一道流泉高墙长廊，将北侧教师原创楼和南侧教学楼合二为一。30°、60°直角三角形平面原创楼轴线网络与东侧校园建筑群居于同一直角坐标系平面，凹弧形西北外墙与北面图书馆遥相呼应。教学楼南侧外墙凸弧形轮廓线与南侧车道平行，东西两侧外墙轮廓延长线指向凸弧形的中心。三部分建筑自然融合成艺海拾贝。

原创构思的进一步发展，是建筑室内环境顺畅的流线、合理的布局、精确的房间定位、行云流水般的空间流、公共聚会地方与最佳艺术品陈列环境：天然高侧光照射的石窟、挂画的壁和摆放雕塑的龛。教学楼四翼建筑围绕室外庭院，教室后墙门式落地长窗采光；三层高门厅北墙顶斜面玻璃天窗引进天然顶侧光，通高落地长窗系列借景高墙流泉长廊。原创楼辟四分之一圆平面玻璃天窗采光中庭，周边教师工作室是沉思的窟，外墙中部深窗加两侧陈列壁龛。厚墙深窗、高深门洞与通天垂直裂隙，涌起建筑巨石雕塑与绝壁洞窟氛围的高潮。

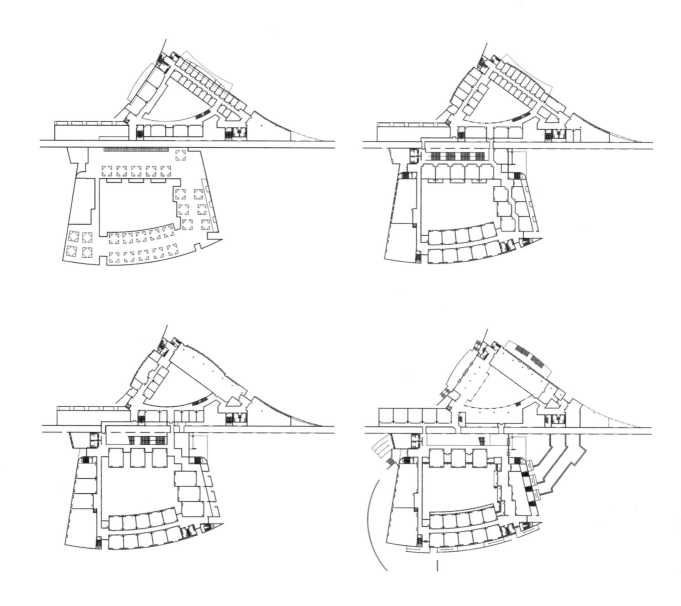

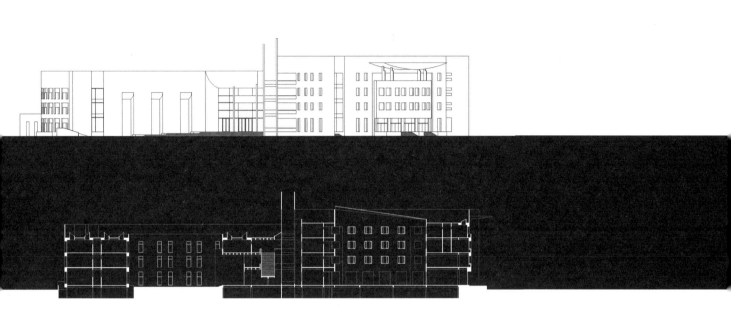
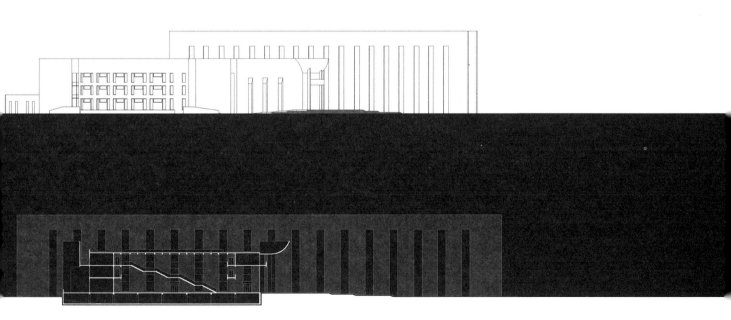

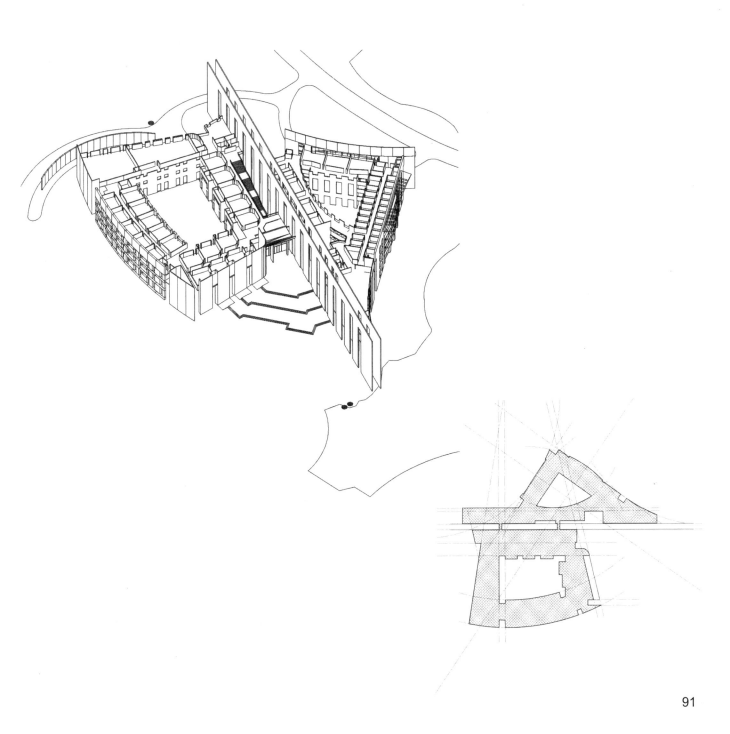

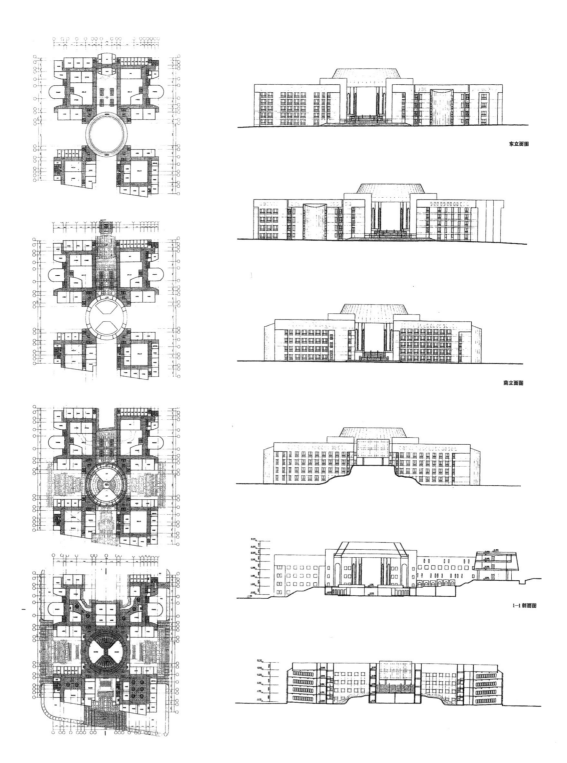

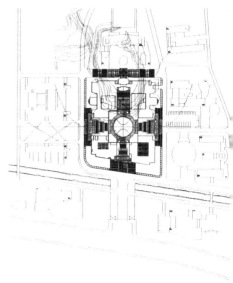

镇江江苏理工大学教学楼建筑设计方案 1999

The Scheme Design of Teaching Building of Jiangsu Technology University.

35 000m²教学综合楼建筑群，得天独厚坐落在长江边丘陵地上，南北中轴线地形纵剖面有两层楼天然高差。基地中央，布置正圆形平面建筑第二层室外入口平台作为聚会与交流中心，与西南、东南、东北、西北四角围绕室外庭院布置的四层高教学楼单元楼梯电梯门厅相连接。三层高办公楼单元位于南北中轴线北端三至五层平面，沿中轴线筑露天步行台阶风景廊道，下行分别与中央主入口平台及南端自然地面平接。东西向校园车行主干道在中央平台下方地面层穿越，中央平台上方第五层标高位置筑正圆环形露天风景走廊，将四组庭院式教学楼单元屋顶花园融为一体；下方八片四层高拱门支撑，构成中央平台上每一个教学楼单元入口门厅的左右拱卫。

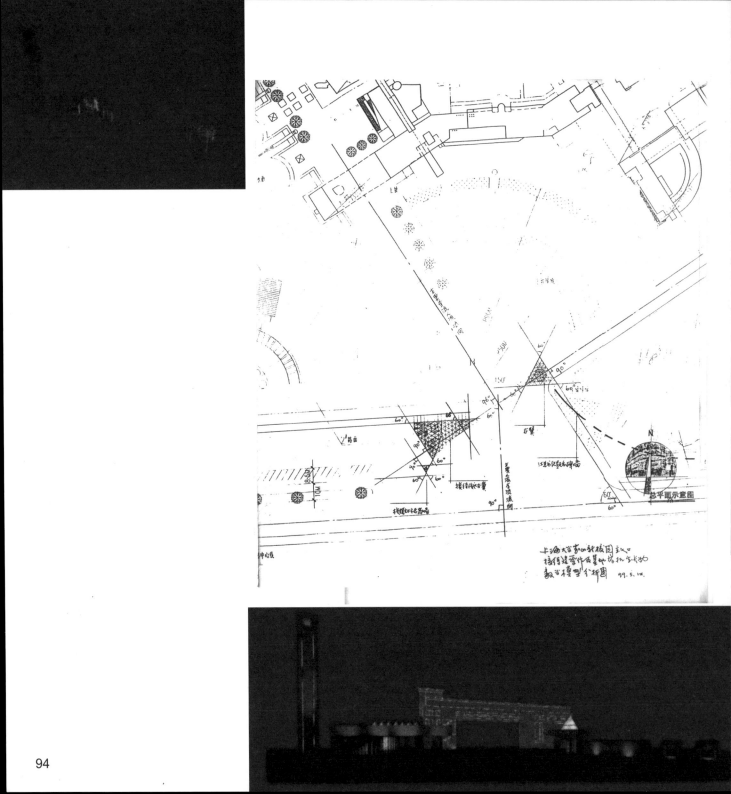

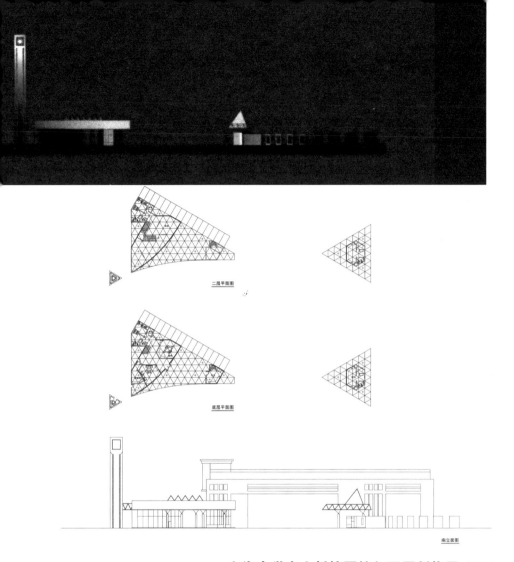

上海大学宝山新校园校门区原创构思 1999

The Concept Design of Entrance, Shanghai University: 30° & 60° triangle plane, entrance plaza, diamond molecular structure.

30°、60°直角三角形平面校前区广场,以天然钻石分子结构空间点阵特征、正三角形平面和正四面体造型为母题,开拓一组雕塑形建筑,酿成强烈的视觉冲击力。校门不设门框,直角三角形长边布置标志高塔与门卫接待左翼建筑,短边右翼建筑、沿四分之一圆周圆弧形校名碑墙和正方框形系列围墙带。36m高正三角形标志塔北面面向校园,西南和东南面向东西向主干道;高塔三个正方形顶面镶嵌校徽灯箱,下方镶嵌西南和东南两条通高400mm宽银灰色镜面玻璃灯光带。校门左右翼建筑,采用正三角形单元的空间、铺地纹样和独立圆柱或轻钢结构的玻璃建筑支撑的金属空间网架系统,上覆正四面体的玻璃天窗。右翼建筑位于从行政办公楼方向来的道路中心线上,左翼建筑接待与传达两部分之间有6m高的敞厅,与从美术学院方向来的人行道融为一体。

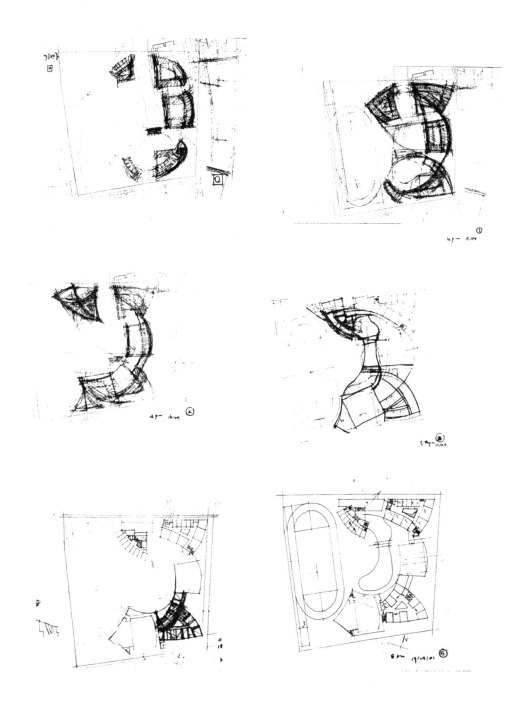

上海音乐学院荟思实验学校建筑设计方案 2001

The Scheme Design of Wisdom School, Shanghai Music Academy: gradually heightened five education buildings enclosing crescent-shaped central lake.

引进上海音乐学院品牌九年一贯制实验学校,致力于以爱和美激发学生的原创思维。总平面布置与建筑风景造型,保留中段河滨构成月牙状白沙底风景水体,连接大片绿地,构成校园读书人视觉和聚会中心。小学部、图书电脑中心、中学部、体育艺术馆和宿舍楼五部分建筑及其室外平台,如众星拱月环绕中心水体布置,二层高中庭主门厅位于东南角。

建筑以室内设计为核心,依据教学活动的策划提供各种功能用房。在保证安全疏散的前提下,提供采光中庭内廊式室内和室外公共活动与陈列空间。微弧形平面的建筑由东南向西北逐级升高,在三维透视层面表现为渐开线上扬形体的动态美。高而窄的框景窗序列及其引发的四黑三白外墙构图键盘特征,表达音乐、数学与建筑的殊途同归:空间高下曲折宛若藏宝洞窟,建筑厚墙深窗,绝壁进退,巨石般造型铸造刚正不阿的师石大器。

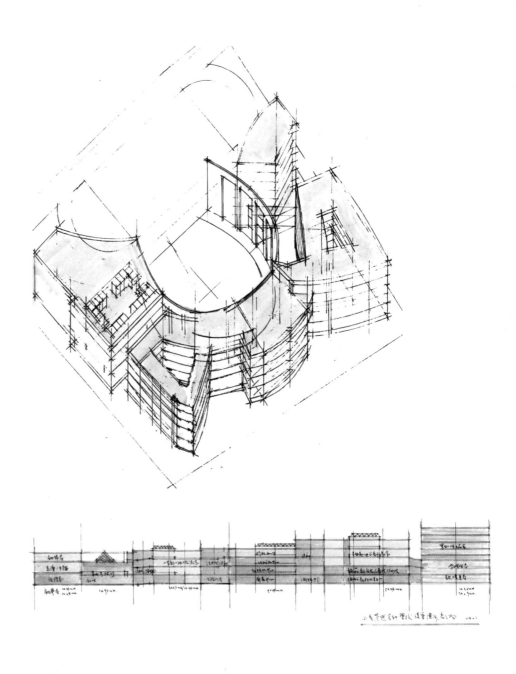

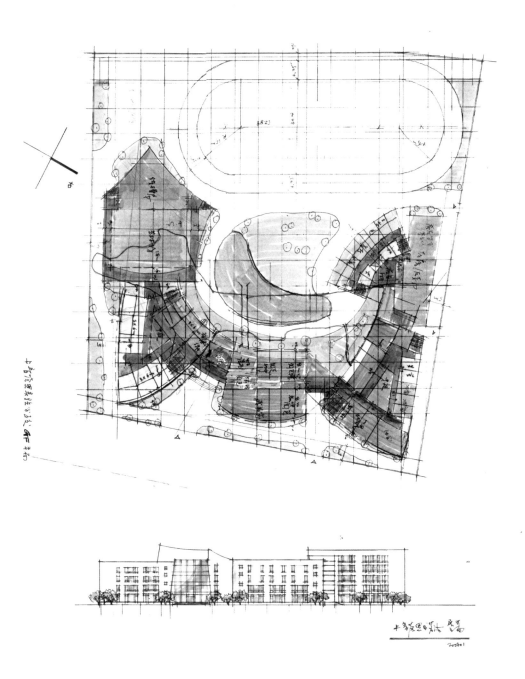

三层平面
上音荟思实验学校校园建筑设计方案
2001.5

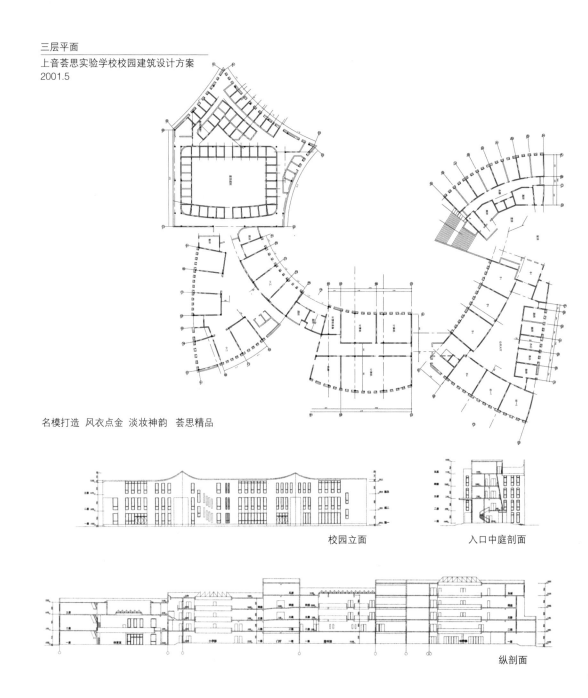

名模打造 风衣点金 淡妆神韵 荟思精品

校园立面

入口中庭剖面

纵剖面

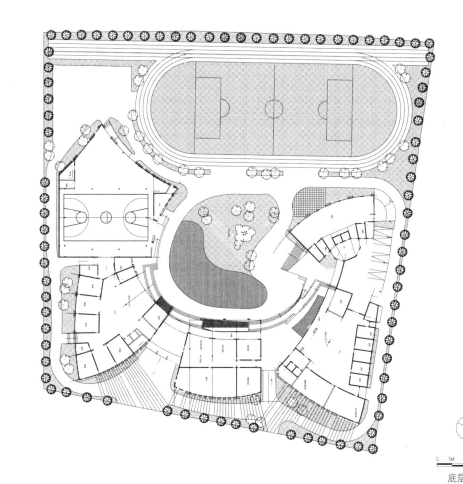

底层平面

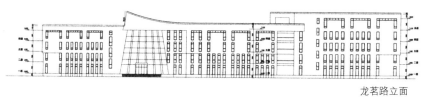

龙茗路立面

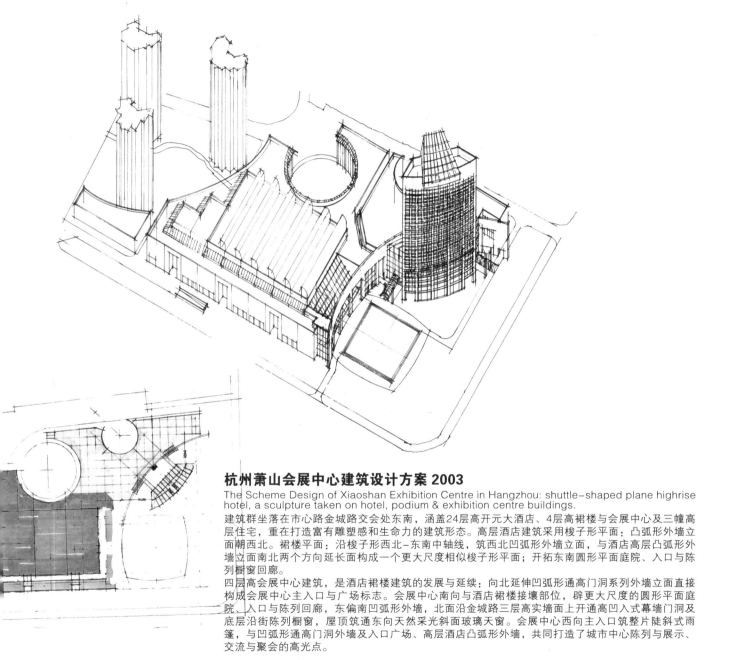

杭州萧山会展中心建筑设计方案 2003

The Scheme Design of Xiaoshan Exhibition Centre in Hangzhou: shuttle-shaped plane highrise hotel, a sculpture taken on hotel, podium & exhibition centre buildings.

建筑群坐落在市心路金城路交会处东南，涵盖24层高开元大酒店、4层高裙楼与会展中心及三幢高层住宅，重在打造富有雕塑感和生命力的建筑形态。高层酒店建筑采用梭子形平面：凸弧形外墙立面朝西北。裙楼平面：沿梭子形西北-东南中轴线，筑西北凹弧形外墙立面，与酒店高层凸弧形外墙立面南北两个方向延长面构成一个更大尺度相似梭子形平面；开拓东南圆形平面庭院、入口与陈列橱窗回廊。

四层高会展中心建筑，是酒店裙楼建筑的发展与延续：向北延伸凹弧形通高门洞系列外墙立面直接构成会展中心主入口与广场标志。会展中心南向与酒店裙楼接壤部位，辟更大尺度的圆形平面庭院、入口与陈列回廊，东偏南凹弧形外墙，北面沿金城路三层高实墙面上开通高凹入式幕墙门洞及底层沿街陈列橱窗，屋顶筑通东向天然采光斜面玻璃天窗。会展中心西向主入口筑整片陡斜式雨篷，与凹弧形通高门洞外墙及入口广场、高层酒店凸弧形外墙，共同打造了城市中心陈列与展示、交流与聚会的高光点。

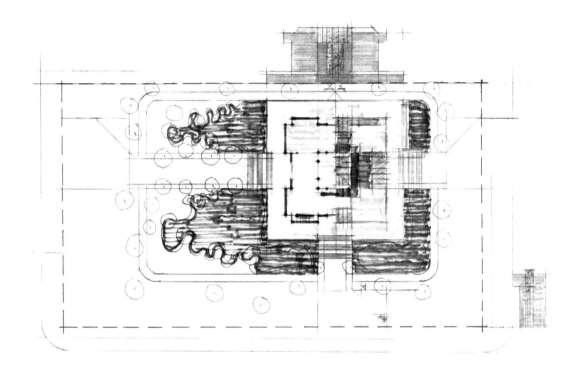

山东曲阜孔子博物馆原创构思 2006
The Scheme Design of Confucius Museum in Qufu: square plane museum, triangle plane subsidiary, central fountain.

由正方形平面4层博物馆主体建筑、等腰直角三角形平面6层西南与东北两座辅楼兼入口建筑组成。博物馆南东北三面临池，从二层平面入口，拥有三层高的室内中庭、落地长窗和转角窗及屋顶层陡斜式玻璃天窗和入口雨篷。辅楼拥有五层高落地长窗和转角窗及屋顶层陡斜式玻璃天窗。水池、玻璃幕墙与实体建筑造型与风景，表现仁者乐山、智者乐水、藏宝洞窟、如切如磋的境界。

城市与国土地方
Urban & Territory Places

房前葡萄架下

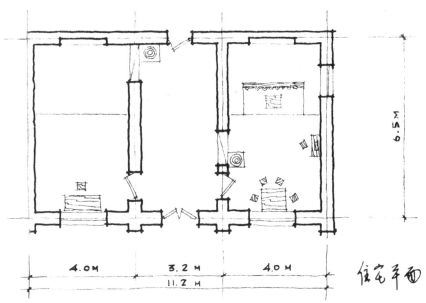

住宅平面

吐鲁番前进大队新农村住宅调查 1978

The Loess Arch Residence in Turupan: desert, gully, greenland village, loess wall & arch farmer residence.

八百里火焰山吐鲁番盆地，广漠戈壁极少植被；而在地貌突然下陷的断崖沟谷中则潺潺泉溪、黄土良田绿野、村居散点其间。维吾尔族农民利用当地干燥无雨特殊气候条件，以黄土构筑厚实墙身与拱券式屋顶、结构简单而隔热性能良好的庭院式住宅。三开间宽扶壁柱土拱窑洞，布置土炕、火墙、土墙壁龛与灶台，拱顶开小窗通风采光。庭院土块花格围墙，拱形院门；前院设葡萄架，后院种植蔬菜果木，大面积为绿化覆盖，形成良好的小气候。夏季吐鲁番大地热浪滚滚，而土拱窑洞内外，则白杨参天、果木掩映，凉气袭人，为人创造了一个休养生息的美好环境。

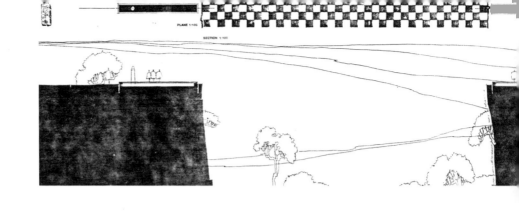

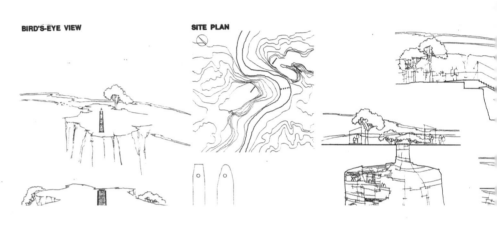

BIRD'S-EYE VIEW　　　SITE PLAN

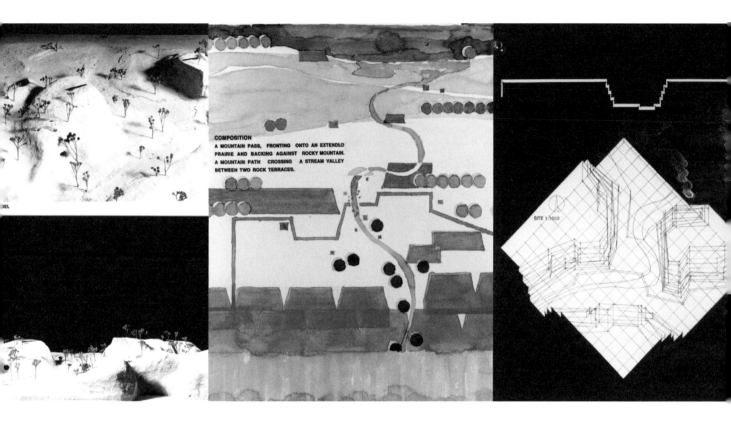

智慧谷 原创桥 1990

A Studio Design in the University of Manitoba-Wisdom Fountain & Creative Bridge:contrast & harmony between the natural & the rational.

在马尼托巴大学建筑学院攻读硕士学位期间课程设计《一座通向自然的桥》，教学指导书源自诺伯格·舒尔茨·克里斯逊著作《地方密码》的一段文字。海德格尔开拓了"桥"的意蕴来诠释琳琅满目的人生大戏："建筑"集聚环境、实施综合体、赋予象征符号，作为视觉环境体验呈现在世人面前。《桥》的设计原创，通过手稿、渲染、数学与实体模型，显现了生存地方感性跋涉与理性探索两种文化体验的切磋与磨合。在山溪两岸绝壁顶，分别筑垂直于山溪直线形抽象流泉及其自谷底登顶两条特征迥异的路径：一经穿越洞窟攀登崎岖山道来到流泉远溪端自然感性圆锥形雕塑，一借透明玻璃顶自动扶梯与平缓芳草地抵达流泉远溪端抽象理性切片状雕塑，智慧泉架起原创桥。

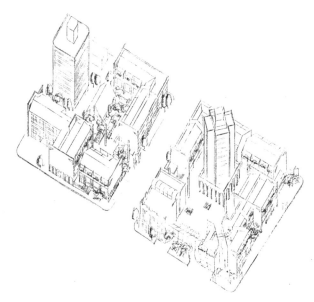

PACIFIC PLACE, AN URBAN DESIGN PROPOSAL FOR WINNIPEG'S CORE AREA
BY SHAOKAI KONG, B.ARCH., M.ARCH., M LAND. ARCH.   DEPARTMENT OF LANDSCAPE ARCHITECTURE   UNIVERSITY OF MANITOBA 1991

加拿大温尼伯中国城城市设计 1989–1992

The Urban Design for Winnipeg City Chinatown: a studio design for Master Degree in UM, preserved forever at the Art Museum of University of Manitoba: to upgrade Chinatown into a urban gathering place, integrating nature, history & frontier of civilization.

马尼托巴大学建筑学院硕士学位毕业设计,马尼托巴大学美术馆永久收藏。原创构思旨在将位于市中心四街区中国城提升为温尼伯城市休闲旅游的高光点。保护传统商业文化交流模式与街景,打造开放式步行街区,在城市主干道 Main Street 与城市中心交会点,辟出与大街成45°角的商业文化风景步行街,将城市人流直接引入城市核心历史博物馆区。在导师莱特和纳尔逊等三位老师指导下,精读与笔记了诺伯格·舒尔茨克里斯逊著作《地方密码》、《安居之道》和《建筑文选》,并将其中提出的主题与背景、认知和识别过程与选择和定位过程和造型体系(形态学)、空间构成(拓扑学)与风景网络(类型学)理念与经验运用于毕业设计的全过程。

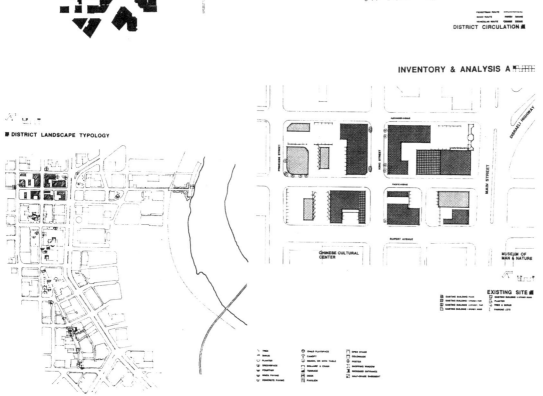

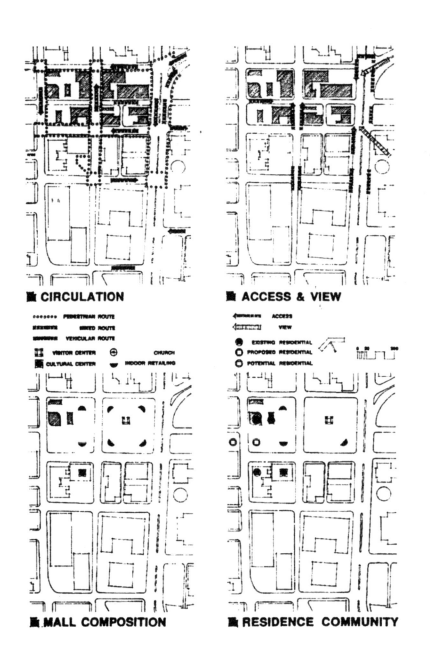
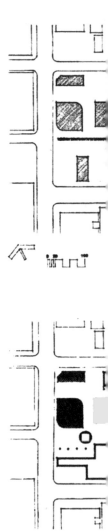

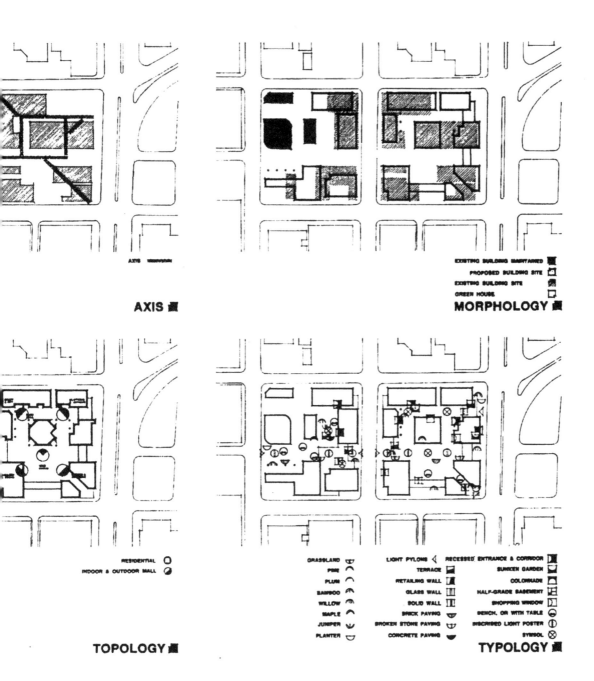

EXISTING EAST-WEST TOPOLOGICAL SECTION, PACIFIC

EXISTING WEST-EAST TOPOLOGICAL SECTION, PACIFIC

EXISTING TOPOLOGY SCALE 1:300

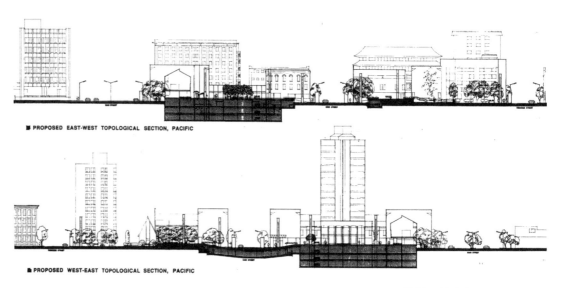

PROPOSED EAST-WEST TOPOLOGICAL SECTION, PACIFIC

PROPOSED WEST-EAST TOPOLOGICAL SECTION, PACIFIC

DESIGN TOPOLOGY SCALE 1:300

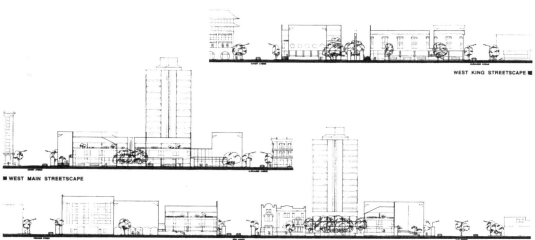

WEST KING STREETSCAPE

WEST MAIN STREETSCAPE

NORTH RUPERT STREETSCAPE

STREETSCAPING SCALE 1:300

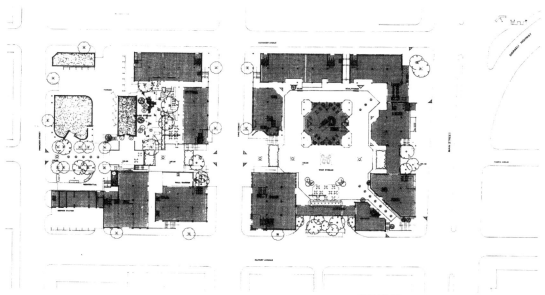

SITE PLAN STREET & PLAZA LEVEL, 100.00 - 105.00

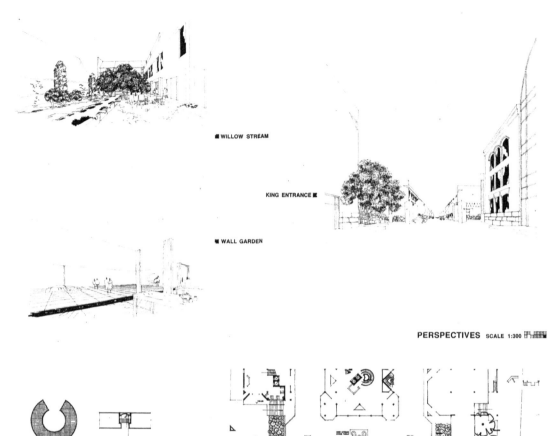

■ WILLOW STREAM

■ KING ENTRANCE

■ WALL GARDEN

PERSPECTIVES SCALE 1:300

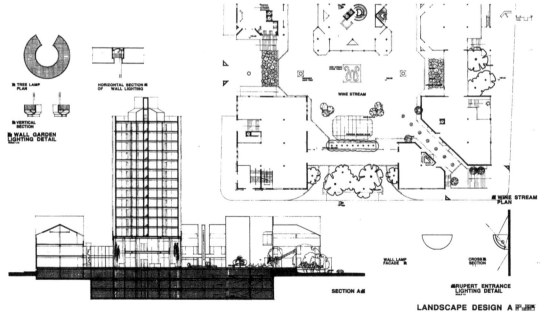

LANDSCAPE DESIGN A

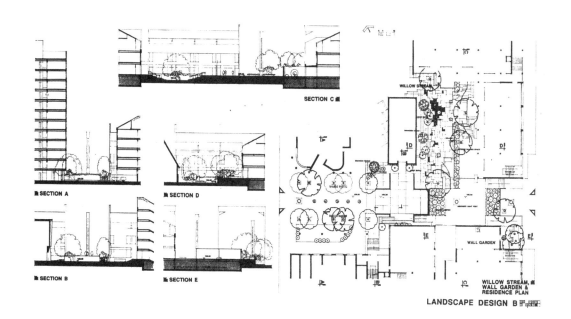

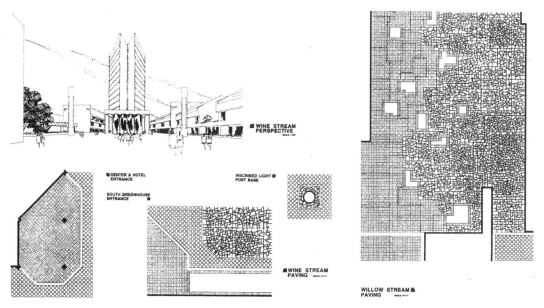

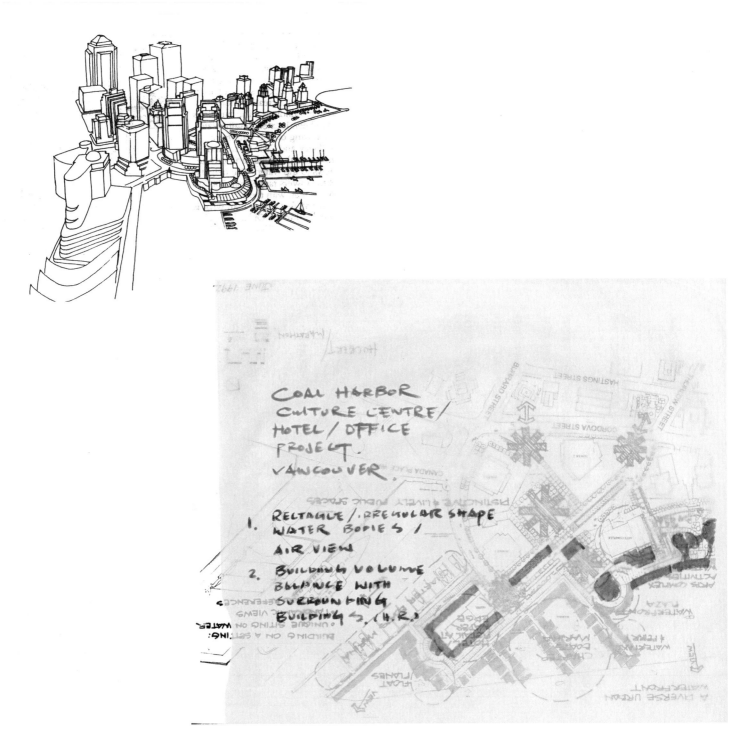

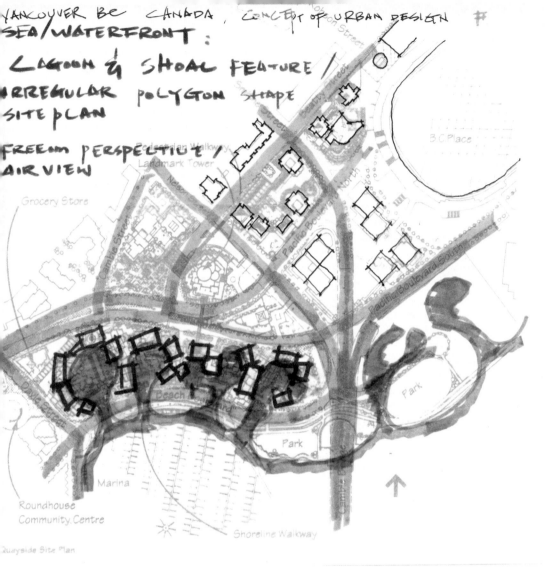

加拿大温哥华海滨地发展原创构思 1992

The Concept for Vancouver City Waterfront Area Development: residential buildings scattering along lagoon-like coastline in False Creek, Coal Harbor roof-top's fountain.

1992年夏，应邀参加小溪（False Creek）、煤码头（Coal Harbor）房地产发展公众听证会，提出了两点策划建议：False Creek 地块，沿天然海岸线开辟新的人工泻湖浅水体，新开发数幢住宅楼沿人工泻湖岸线作自然式总平面布置；在Coal Harbor 商业文化中心多层建筑物屋顶做浅水屋面，隔热保温并构成水风景；居住和公共建筑与海景融为一体，取得土地发展的最佳生态、风景与楼面价值。

绍兴古越龙山城市广场原创思溯源影像
P122绍兴采石石宕遗址 左上吼山烟萝洞
左下柯岩云骨 右下吼山棋盘石 左上石佛寺
P123左下夕阳溪桥 左下一方水土 右下水城人家 右上水乡波月

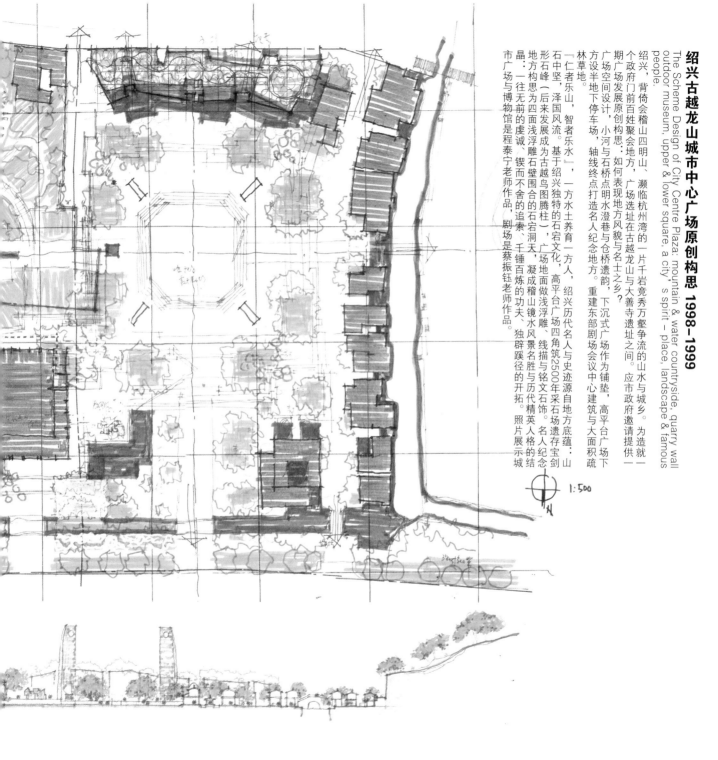

绍兴古越龙山城市中心广场原创构思 1998-1999

The Scheme Design of City Centre Plaza: mountain & water countryside, quarry wall outdoor museum, upper & lower square, a city's spirit – place, landscape & famous people.

绍兴，背倚会稽山四明山，濒临杭州湾的一片千岩竞秀万壑争流的山水与城乡。为造就一个政府门前百姓聚会地方，广场选址在古越龙山与大善寺遗址之间。应市政府邀请提供一期广场发展原创构思：如何表现地方风貌与名士之乡？一期广场空间设计，小河与石桥点明水澄巷与仓桥遗韵，下沉式广场作为铺垫，高平台广场下方设半地下停车场，轴线终点打造名人纪念地方。

"仁者乐山，智者乐水"，一方水土养育一方人，绍兴历代名人与史迹源自地方底蕴：山石中坚，泽国风流。基于绍兴独特的石宕文化，高平台广场四角筑2500年采石场遗存宝剑形石峰（后来发展成为古越鸟图腾柱），广场地面做浅浮雕，线描与铭文石饰。名人纪念地方构思为四面浅浮雕石壁围合的石宕洞天，凝成稽山镜水风景名胜与历代精英人格的结晶：一往无前的虔诚、锲而不舍的追索、千锤百炼的功夫，独辟蹊径的开拓。照片展示城市广场与博物馆是程泰宁老师作品，剧场是蔡振钰老师作品。

1:500

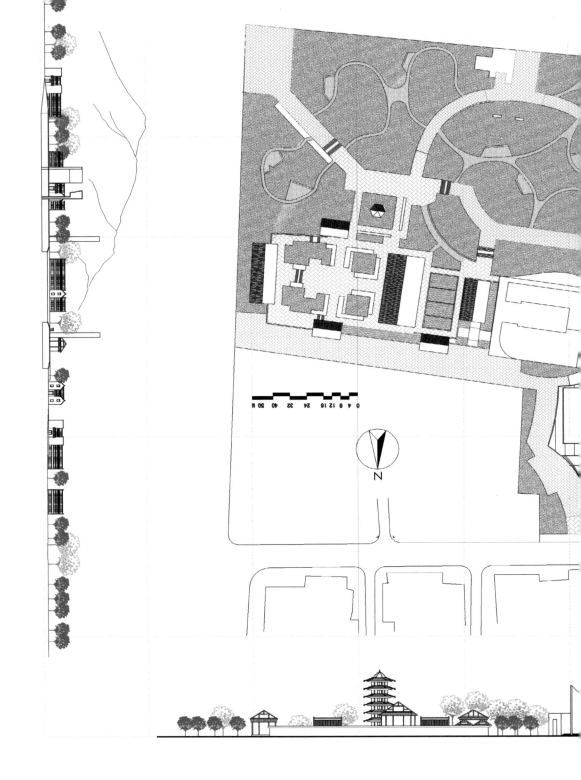

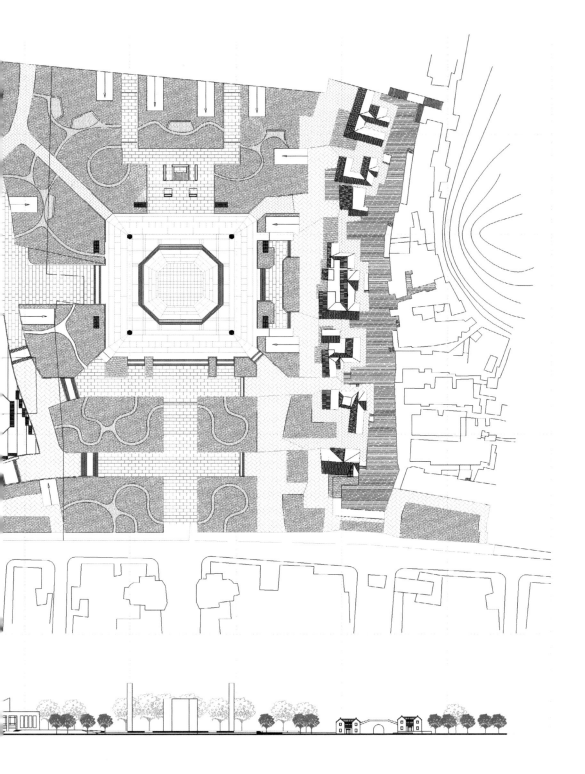

千名竞秀 鉴万壑争流 百代楼稼尽在 长城

129

《包头北梁晋风古城保护与发展》照片图片注释索引

（页面图照注释顺时针为序）

P136 自然历史底蕴：南海子+黄河，23km² 黄河湿地，黄河+湿地，8 000亩黄河古道南海子

P137 自然历史底蕴：古城鸟瞰，东河槽，阴山山脉，同治太平桥

P138 建筑与文化：晋剧演员水上漂王玉山演出剧照，二人台，二人台，三三摄影社

P139 建筑与文化：1927斯坦因无声电影画面：王府摔跤，王妃与王子，城门，城墙

P140 宗教建筑：吕祖庙大殿，清真西寺，官井梁基督堂，清真大寺

P141 宗教建筑：天主教堂，东河槽龙泉寺，300年藏传佛教福徽寺，新太店巷居士林

P142 晋风民居四合院：风雪王国秀1号院墙街巷，玄关照壁，鸽子和为贵，五原厅

P143 晋风民居四合院：照壁庭院，丰备仓益丰堂拱门，王国秀1号庭院，古树庭院

P144 古城街景：郭家巷，通向吕祖庙街巷，200年财神庙广场古戏台，坡地宅院

P145 古城街景：蒙古族白家大院，三官巷二道巷，新太店巷乔家丝绸店，商会大门

P146 庙堂风貌：福徽寺大殿，福徽寺中庭回廊藻井，清真西寺大殿，清真大寺大殿

P147 宅院风景：红楼梦白蛇传彩画额枋木隔断，供奉，彩画炕围子，炕上柜

P148 建筑细部：砖雕额枋门饰，砖雕额枋门框，砖雕飞檐山墙，砖雕飞檐

P149 四合院拱门匾额："莺歌燕舞"，"铭绣春"，"诗礼传家"，"平为福"

包头北梁晋风古城保护与发展　2003-2006

*包头北梁晋风古城自然与建筑遗存调查　　　　　　　　　　　　*东河沿河土地城市发展原创构思
*包头北梁九江口核心区规划与城市设计　　　　　　　　　　　　*包头东河行政办公楼建筑方案设计
*九江口核心区14#地块规划与城市设计　　　　　　　　　　　　*传统晋风民居四合院优化发展原创构思
*包头东河水风景环境优化原创构思　　　　　　　　　　　　　　*北梁两百年九江口城市广场重建设计

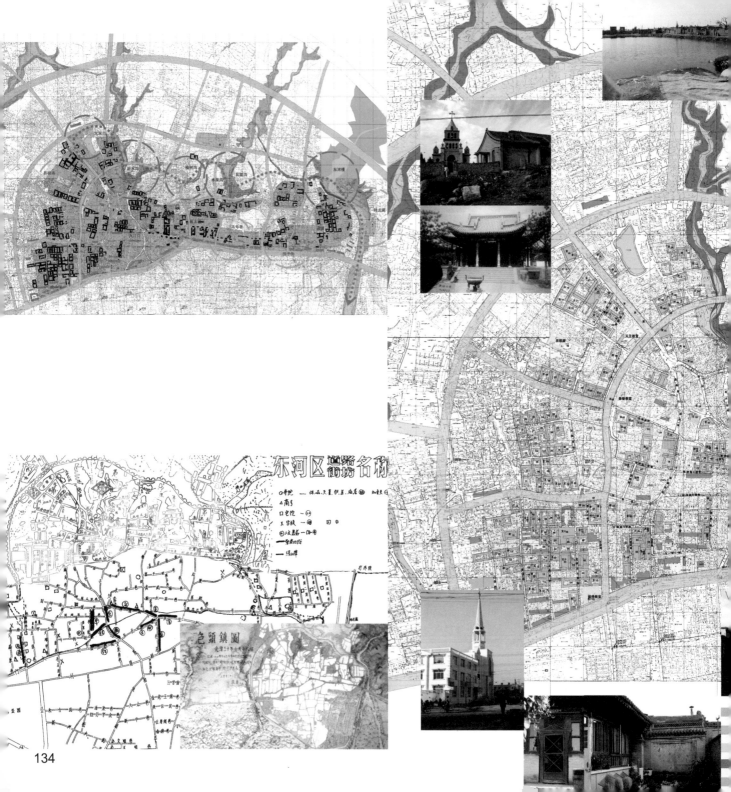

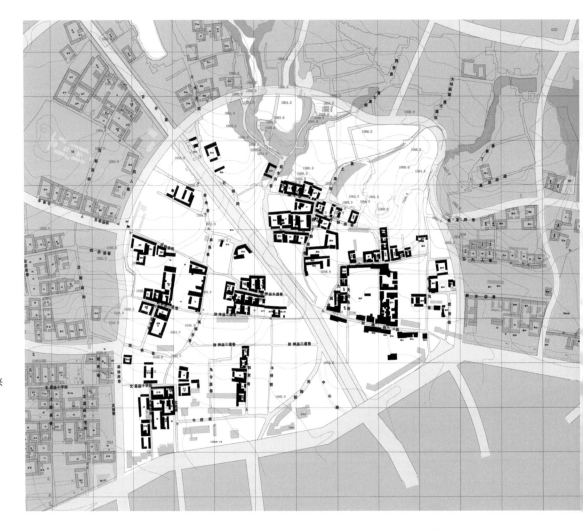

九江口城市复兴
背景环境2004

保护自然　　保护历史
珍惜　每一寸国土环境
创造建筑城市风景精品

The Urban Design for Jiujiangkou Core Area: 200-year old central plaza, streets, shops, residence & urban environment.

晋风古城中心九江口核心区占地约0.49km²，区内古街巷古宅古院轮廓尚存，一条30m宽的新建混凝土马路将古城中心财神庙广场劈成两半。九江口核心区古城保护与发展原创构思定位在：继承历史文脉，将自然与建筑遗产保护与发展作为主题；用地划分为东北台地别墅区、西南商业居住区和两者之间财神庙中心区三大块。

以财神庙、古戏台和广场为交流与聚会中心，化解混凝土马路，周围引进室内步行街、室内购物中心、室内贸易市场和九江口文化艺术中心，并辐射出三条轴线：西向妙法寺开通庙前街，南向连接和平路中轴线，东向新太店巷开通西新太店巷。区内路网保存4~9m宽100m左右间距古街巷，局部拓宽至14m宽，人车并流；保留晋风民居商号作坊宅和古街巷格局，局部加玻璃连廊或天窗开拓为作坊、陈列馆、艺术家工作室或商务办公特色街区，创建新的二至四层庭院式居住街坊。与包头城市规划设计院合作规划与设计。

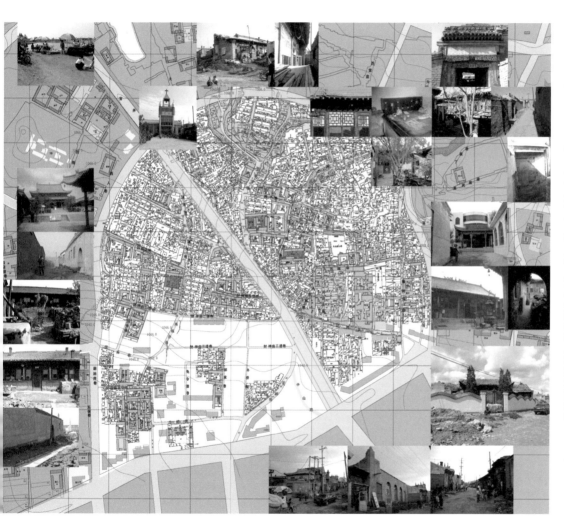

包头北梁九江口核心区规划与城市设计 2003-2004

龙藏新城九江口控制性规划
2003. 8 - 2004. 2
大器建筑设计咨询(上海)有限公司
包头市　城市　规划　设计　研究院

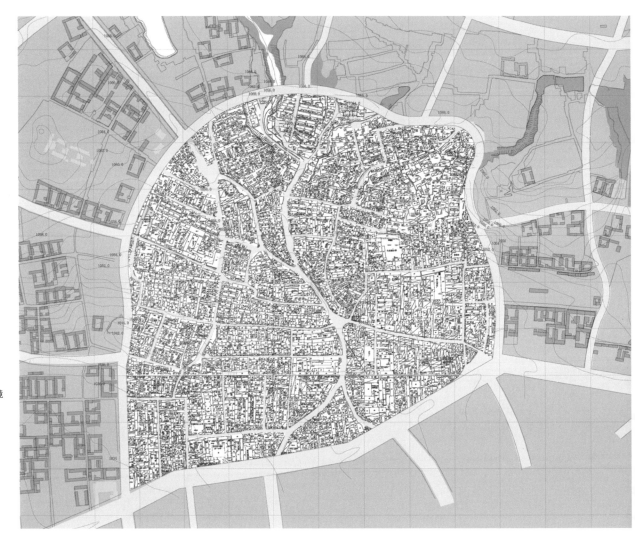

九江口城市环境
1900'S

九江口航测照片
2004

主题与背景：大面积待拆迁开发的土地作为背景，列入保护范围的自然与建筑遗产作为城市复兴主题。

轴线：以财神庙、戏台和广场为中心，三条轴线分别向西连接妙法寺庙前街，向南连接和平路中轴线，向东连接新太店巷中轴线。

人流：延续古城小街闹市、廊庙接踵、熙熙攘攘的人流特征，大部分街巷人车并行，沿三条轴线步行为主。东北台地和西南商住区引进现代车行交通系统。

入口与景向：入口从东南西北四个方向进入，财神庙广场九条街道会聚：1.妙法寺庙；2.吕祖庙街+财神庙头道巷；3.财神庙二道巷；4.财神庙三道巷+光辉街+中山路；5.财神庙南街；6.财神庙东街；7.西新太店巷；8.久长城巷；9.园子巷。
主要景向沿着妙法寺庙前街、和平路和西新太店巷三条中轴线。

九江口
城市复兴结构分析

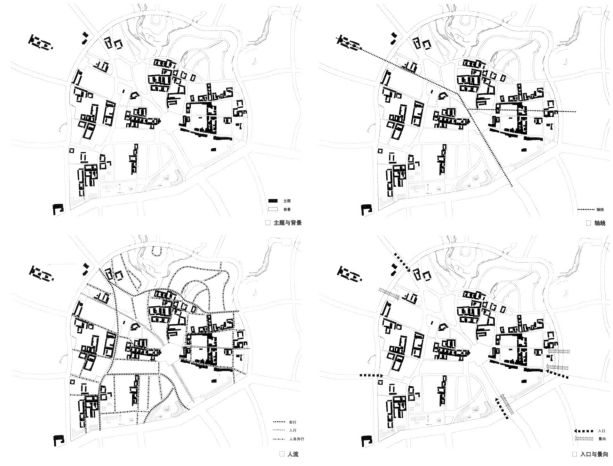

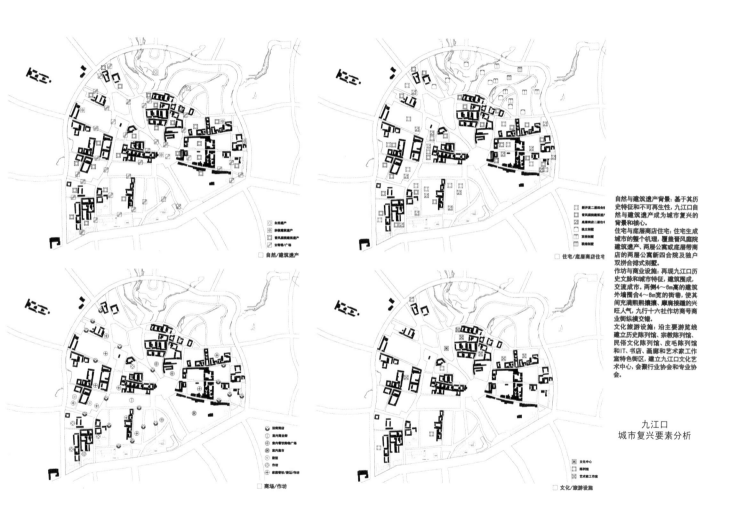

自然与建筑遗产背景：基于其历史特征和不可再生性，九江口自然与建筑遗产成为城市复兴的背景和核心。

住宅与底屋商店住宅：住宅生成城市的整个机理，覆盖晋风庭院建筑遗产、两层公寓或底屋带商店的两层公寓新四合院及独户双拼合排式别墅。

作坊与商业设施：再现九江口历史文脉和城市特征，建筑围成，交流成市。两侧4～8m高的建筑外墙围合4～8m宽的街巷，使其间充满熙熙攘攘、廊屋接廊屋的兴旺人气，九行十六社作坊商号商业街纵横交错。

文化旅游设施：沿主要游览线建立历史陈列馆、宗教陈列馆、民俗文化陈列馆、皮毛陈列馆和IT、书店、画廊和艺术家工作室特色街区，建立九江口文化艺术中心，会聚行业协会和专业协会。

九江口
城市复兴要素分析

整个九江口核心区用地划分为三大块：东北台地别墅区覆盖后水沟台地和榆树沟台地；西南商业居住区北缘至财神庙二道巷、东缘至原北大街；东北台地别墅区与西南商业居住区之间的区域为财神庙中心区。

功能用地规划

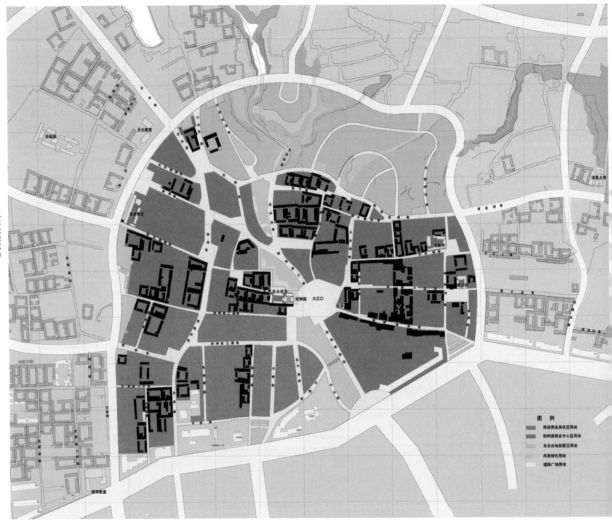

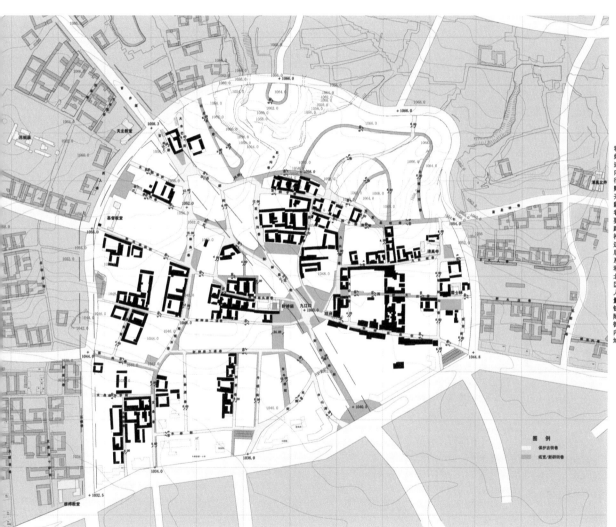

利用原规划已拆迁的五保巷至妙法寺的路基，开通与榆树沟东连通的车行内环线。将北大街环内段分段改造成为建筑用地，步行街和广场。

财神庙中心区保护古街巷格局，开通妙法寺庙前街和向东接通新太店巷的西新太店巷。古街巷人车并行，车辆单向行驶，主要游览线步行为主，庙会期间封路。大西南部商业居住区新辟牛桥街、文曲巷、三道巷国秀车行道系统，在东北台地别墅区新辟后水沟巷、北榆树沟巷车行道系统；并拓宽麻角巷、两眼井巷和大顺恒巷作为车行道贯通南北。

区内 4~9m宽的道路，一般不设人行道。鉴于地形坡陡及年降雨量少，可以用与道路垂直相交的铸铁箅子截留并排除地面降水。除地面停车外，财神庙室内购物街、财神庙室内菜市、财神庙室内购物中心和仓储式超市辟半地下停车场。

交通/规划路网

以保护后水沟、榆树沟及大水巴自然遗产为基础，建设沿后水沟榆树沟经财神庙广场至长胜街和平路口的中心绿地和一些地形剧烈变化部位的集中绿地。
根据交通与路网规划的总体布局，因地制宜设置街头绿地和行道树。
对建筑庭院与沿街建筑宅旁绿地的设计、种护和养护，按自扫门前雪的公益原则进行责任立法，使除了大型公共绿地之外的所有公共风景绿化环境的建设和保护，成为全社会的共同兴趣和责任。

风景绿化规划

1

2

3
4

1. 在中山路和原北大街之间辟40m宽中心绿地，利用原北大街路基上建180m长财神庙室内集市，将长胜街和平路口作为整个九江口核心区最重要的入口方向。确定财神庙广场形状和尺度。

2. 树立西南商业居住区和财神庙中心区建筑总平面框架，确定光辉街室内餐饮广场和财神庙中心区建筑总平面框架，确定光辉街室内餐饮广场和财神庙室内购物广场形状和尺度。

3. 树立东北台地别墅区建筑总平面框架，深化西南商业居住区和财神庙中心区建筑总平面。确定西北仓储式超市形状和尺度。提出妙法寺庙前街与长胜街和平路主入口中轴线交点作为九江口标志性建筑坐落位置。

4. 优化三个分区建筑总平面。在标志性建筑位置建筑58m高跌式标志性建筑九江口文化中心，顶部群360°观景厅，作为九江口的制高点和视线焦点，在主入口中轴线上发展出180m长玻璃柱廊和屋顶玻璃天窗、室内向财神庙集市四层跌落的聚会厅。

建筑总平面布置
原创设计构思过程

西南部商业居住区和财神庙中心区保护古街巷格局和有价值的晋风民居宅院、作坊和商号，新建带底层商店二层公寓四合院。
东北部地区拟建独户、双拼或联排式别墅。

居住与商住规划

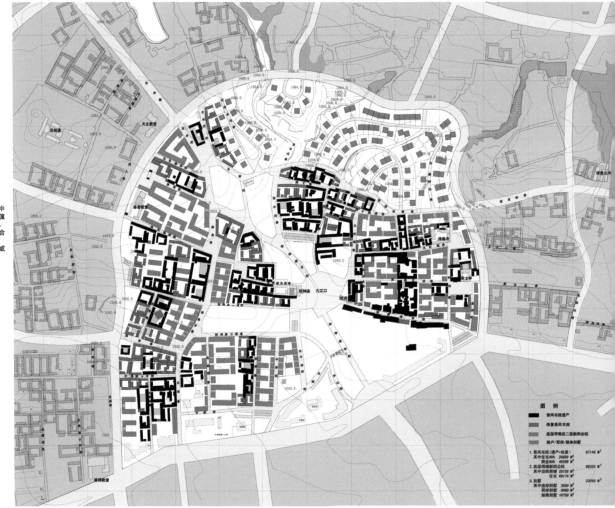

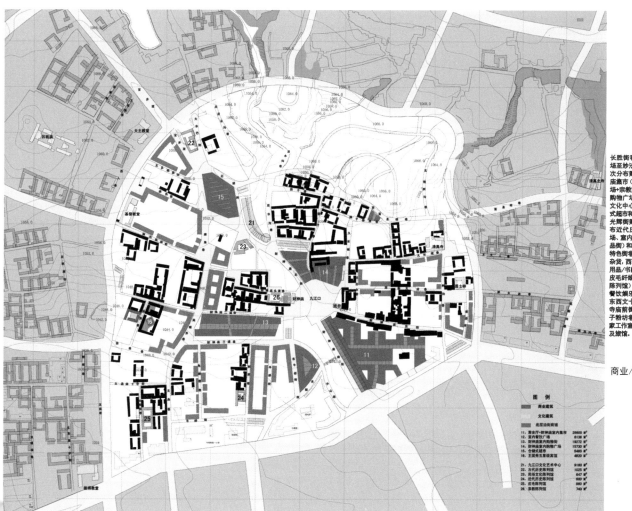

长胜街和平路主入口经财神庙广场至妙法寺山门一线，从南到北依次分布财神庙南街+聚会厅+财神庙集市（小商品+建材）、财神庙+宗教陈列馆、财神庙北街+室内购物广场（综合性百货）、九江口文化中心+民俗文化陈列馆、仓储式超市和古代历史陈列馆。

光辉街财神庙二道巷一线沿线分布近代历史陈列馆、室内餐饮广场、室内步行购物街（国际一流名品街）和王国秀五星级宾馆。

特色街巷与街区：财神庙东街土产杂货，西新太店巷久长城巷IT/文化用品/书籍/画廊，大顺恒巷文曲巷皮毛纤维成衣商店加作坊（含皮毛陈列馆），龟龄酒巷郭家巷牛桥街餐饮娱乐（含近代历史陈列馆），东西文十字街东西吕祖庙街妙法寺庙前街文艺品+餐饮，园子巷图子勃坊巷久长城巷商务办公/艺术家工作室，财神庙二道巷家庭餐饮及旅馆。

商业/文化与旅游规划

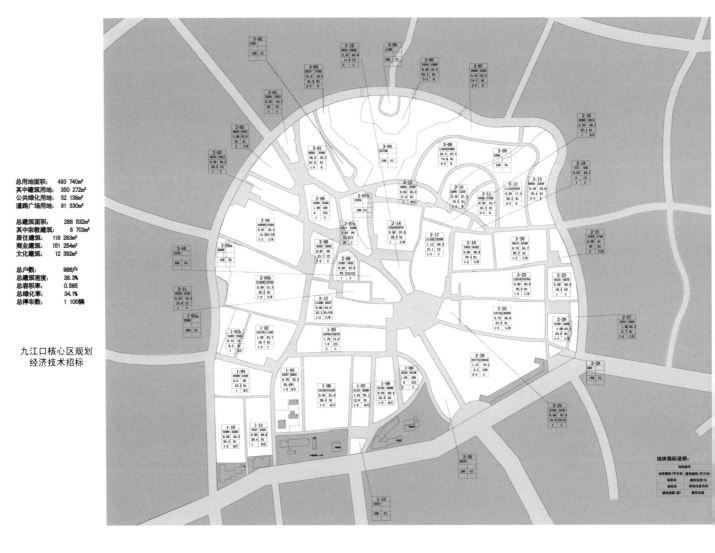

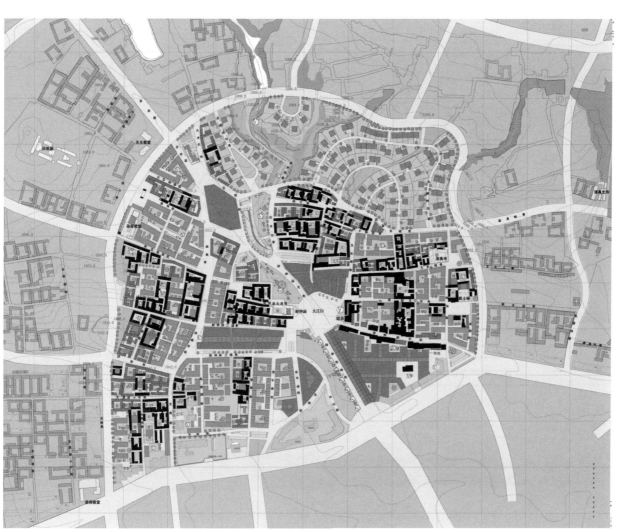

建筑总平面布置

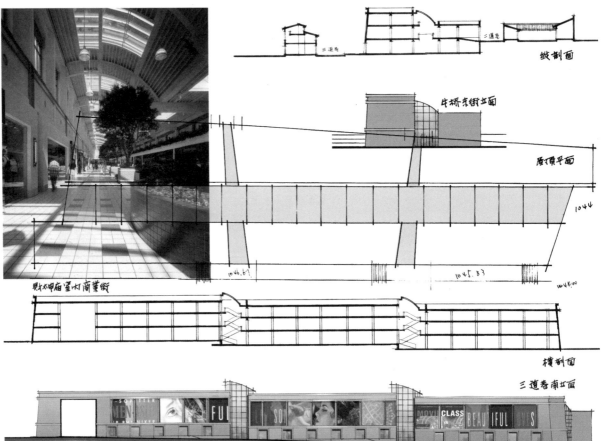

财神庙室内购物街旨在塑造世界一流的高尊地区名品购物街。利用地形高差，二道巷西段在二层楼面进入，三道巷从地面层进入。基于外墙高度与街道路宽的最佳比例，二道巷一侧建三层楼面，三道巷一侧筑两层楼面，形成东西向180m长拥有玻璃卷棚屋顶天窗特色中庭。

财神庙室内商业街
原创设计构思

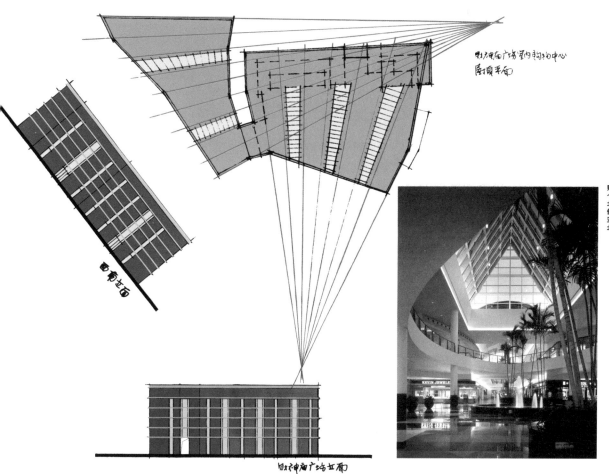

财神庙室内购物广场横跨四子巷两个街区,是坐落在财神庙中心广场北端最重要的城市风景建筑。其南侧16.8m高土红色面砖饰面加水平向亚光不锈钢带的外墙,将构成广场北面的天幕。

财神庙室内购物中心
原创设计构思

九江口文化艺术中心定位在多元文化的融合:古今中外,五洲四海,中西合璧,兼容并包。其定位的一个方面是希望将北梁地区高涨的朝山进香活动引导到天下为公和现代文明的高度。建筑抛物线形平面是基地有机生长物,源自黄河冰凌奇峰突起、黄河鲤鱼跃飞龙门的原创构思。铸成土红色鱼形建筑形体,鱼鳞鱼翅状镜面玻璃饰面抉摇直上360°视野圆形平面屋顶观景厅。

九江口文化艺术中心
原创设计构思

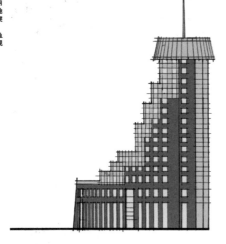

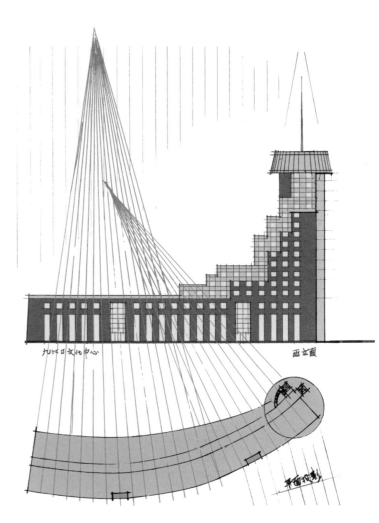

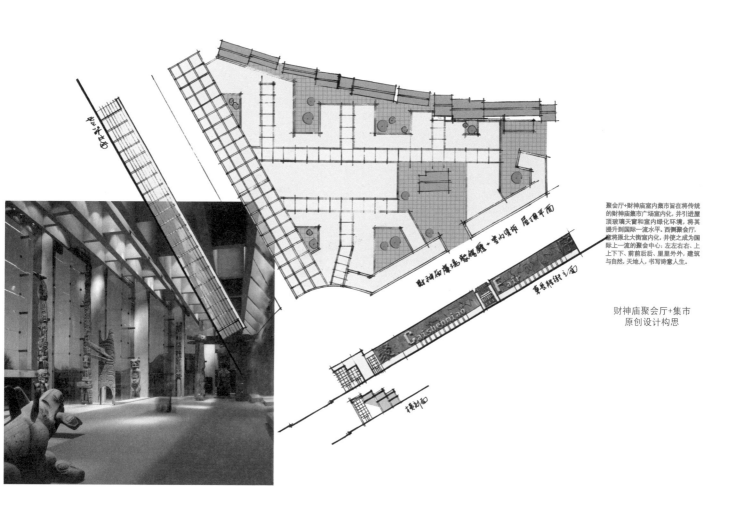

聚会厅+财神庙室内集市旨在将传统的财神庙集市广场室内化,并引进屋顶玻璃天窗和室内绿化环境,将其提升到国际一流水平。西侧聚会厅,拟将原北大街室内化,并使之成为国际上一流的聚会中心:左左右右、上上下下、前前后后、里里外外,建筑与自然,天地人,书写诗意人生。

**财神庙聚会厅+集市
原创设计构思**

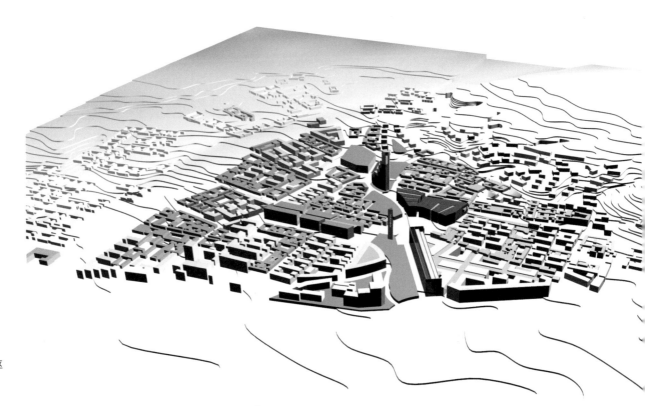

九江口核心区
总体鸟瞰

财神庙聚会厅
室内透视

保护自然　保护历史
珍惜　每一寸国土环境
创造建筑城市风景精品

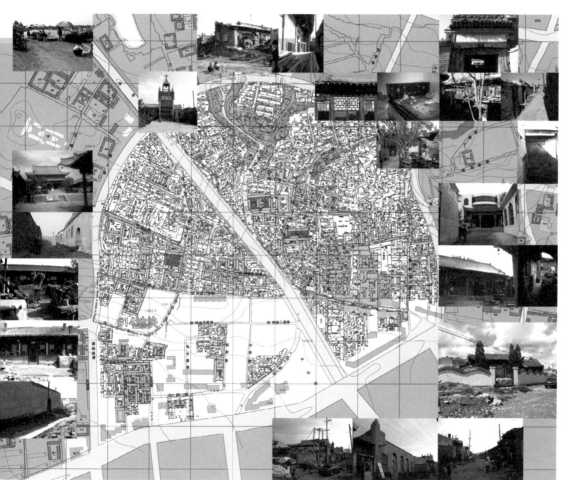

九江口核心区14#地块规划与城市设计2004-2005

The Urban Design for No.14 Plot: east area residence & street-front shops, west area conference center & hotel, based on 150-year old Jin-style residence.

14#地块居于九江口核心区西南部，占地16 242m²。东部居住街坊建筑、道路与绿化采用自然式总平面布置，按日照间距将三四层高的住宅建筑放置在从南到北标高逐渐递升的地基上，造就各自庭院空间。青瓦坡屋顶与高墙深窗的沿街商铺建筑7m高青砖饰面外墙，重在沿街橱窗与陈列的视觉艺术设计。

西部会谈中心宾馆设计，保护与修复四合院民居遗存辟为高档庭院式客房单元，嵌入六层高正方形平面藏召庙厚墙深窗客房主楼及南侧健身中心建筑。国秀巷西侧宴会厅与会议中心新建筑，地面层设两层高中庭多功能大厅；室内北端主墙面饰以《黄河十四走》中国红背景金色龙纹图样，南端水庭院南墙镂刻阴山岩画照壁。沿国秀巷民居宅院7m高外墙大部分保持完好，沿街两层高新建筑稻草黄色面砖外墙、青瓦屋顶加局部玻璃天窗。

龙藏新城九江口控制性规划
2003.8 – 2004.2
大器建筑设计咨询（上海）有限公司
包头市　城市　规划　设计　研究院

171

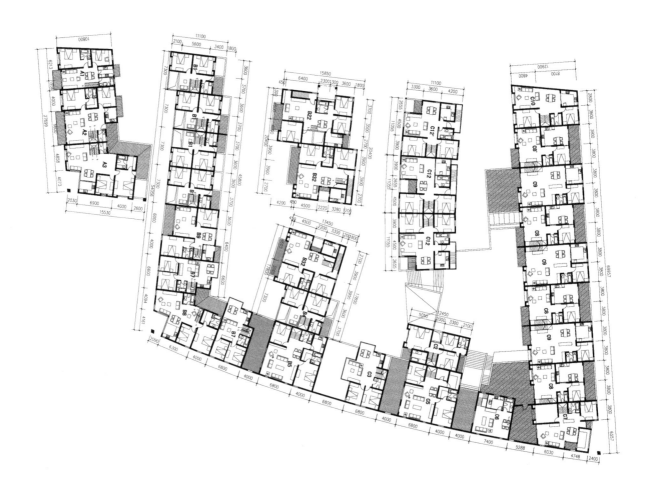

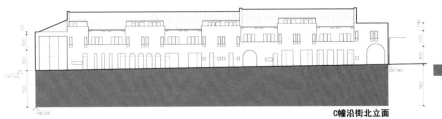

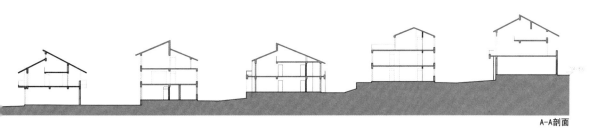

A-A剖面

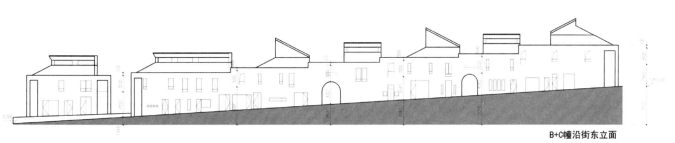

B+C幢沿街东立面

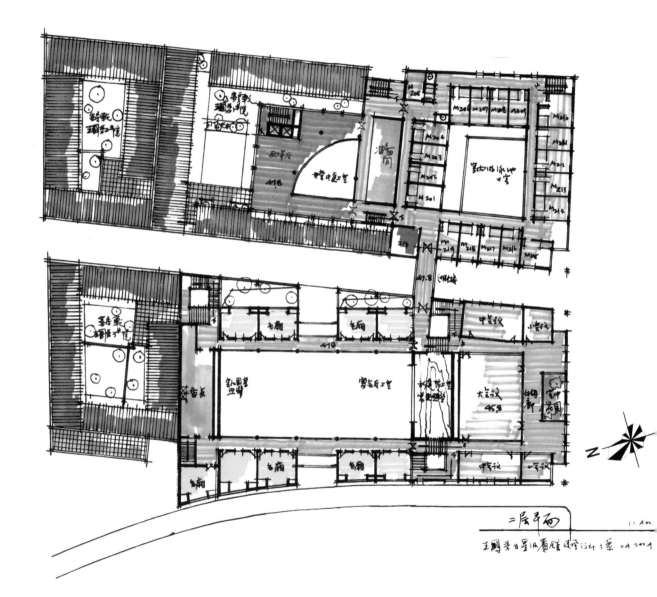

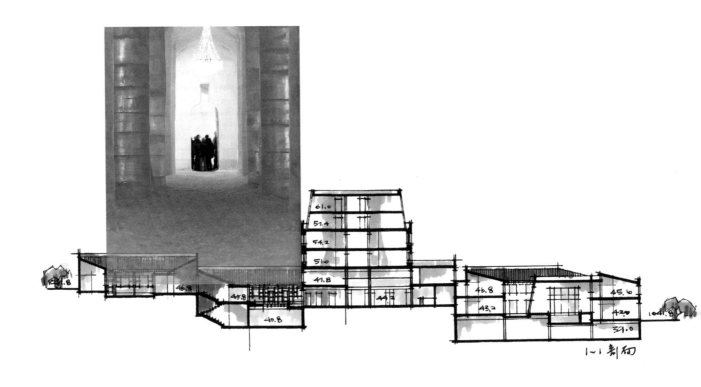

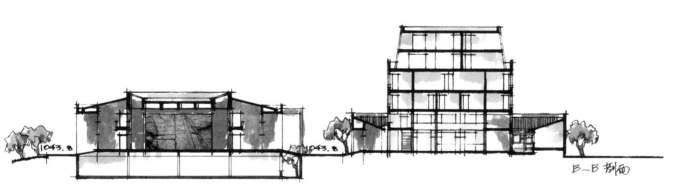

A-A剖面 B-B剖面 1:400

王国秀召星旧宾馆建筑设计方案 04 2004

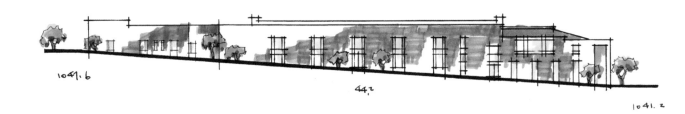

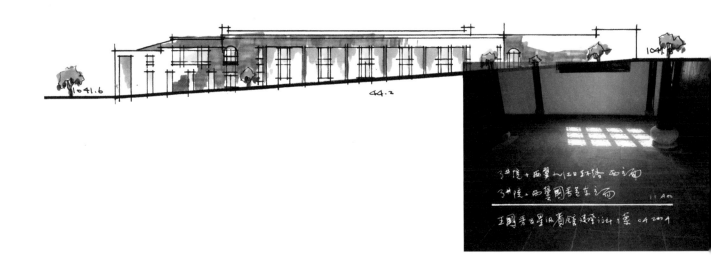

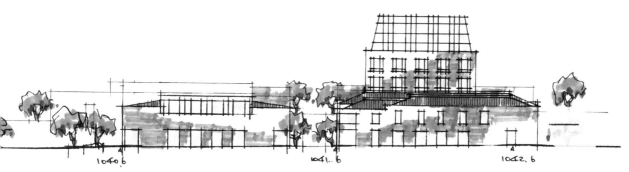

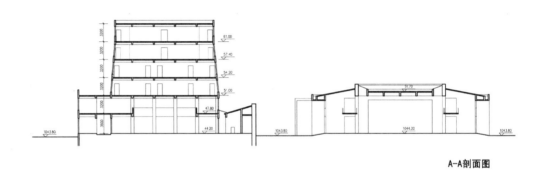

A-A剖面图

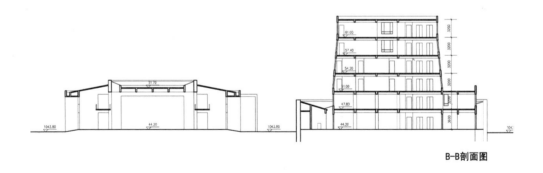

B-B剖面图

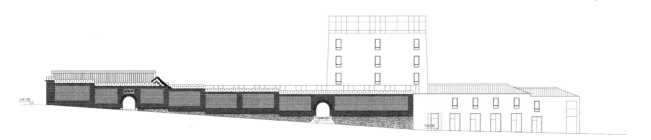

1号2号院西翼+客房楼/健身中心国秀巷西立面

会议中心+宴会厅/客房楼+健身中心国秀巷南立面

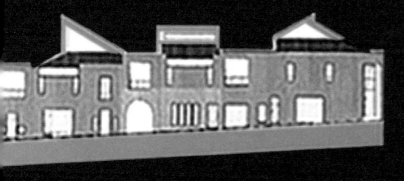

△ A/B/C
东立面

◁ C
北立面

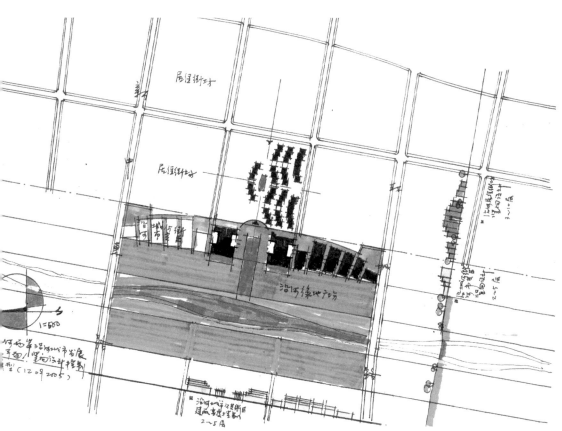

保护自然　保护历史
珍惜　每一寸国土环境
创造建筑城市风景精品

包头东河水风景环境优化
原创构思 2003-2005

A Concept to Upgrade East River's Waterscape Environment: natural river waterscape, scale & investment for man-made waterscape construction.

东河鸟瞰呈现为两道笔直防洪墙夹峙河滩湿地与曲折溪流。2004年初会同湿地专家陆健健老师和水利专家宋德楠老师提出优化策划意见：保护河滩湿地与天然水风景，大幅度降低人工水风景与绿地工程规模与造价。

保留原有防洪墙，新水景工程按20年洪水位标准设计；风景水面横剖面尽量采用缓坡绿地与自然土河床，水面最宽不超过55m；水面纵剖面在地势平缓的北段与南段优化为自然型曲溪与多级低平浅水湖面，减少橡皮高坝与蓄水总量。特别建议将两岸车行道路从河边外移200m，大幅度提升沿河发展地块土地价值。

东河沿河土地城市发展环
境策划意见 2004

An Outline for East River's Waterfront Land Development: invited by the government after the waterscape upgrading development.

2005年东河水风景绿地游览带建设初见成效，每晚数万人蜂拥跃东河两岸。应东河区政府邀请提出东河沿河土地发展策划意见：适应百姓对于交流与聚会的需求，提供各种城市公共服务与设施，促使游览人流回旋活动在东河沿河地带，酿成百业兴旺的城市公共环境。沿河土地使用，应当在水风景绿地游览带与居住街坊之间开辟东西向30～80m²宽沿河公共街区。两岸土地城市环境发展控制内容涵盖沿河土地使用与交通组织、公共街区与居住街区的控制性规划及建筑、风景与室内设计。

龙藏新城九江口控制性规划
2003.8 – 2004.2
大器建筑设计咨询(上海)有限公司
包头市　城市　规划　设计　研究院

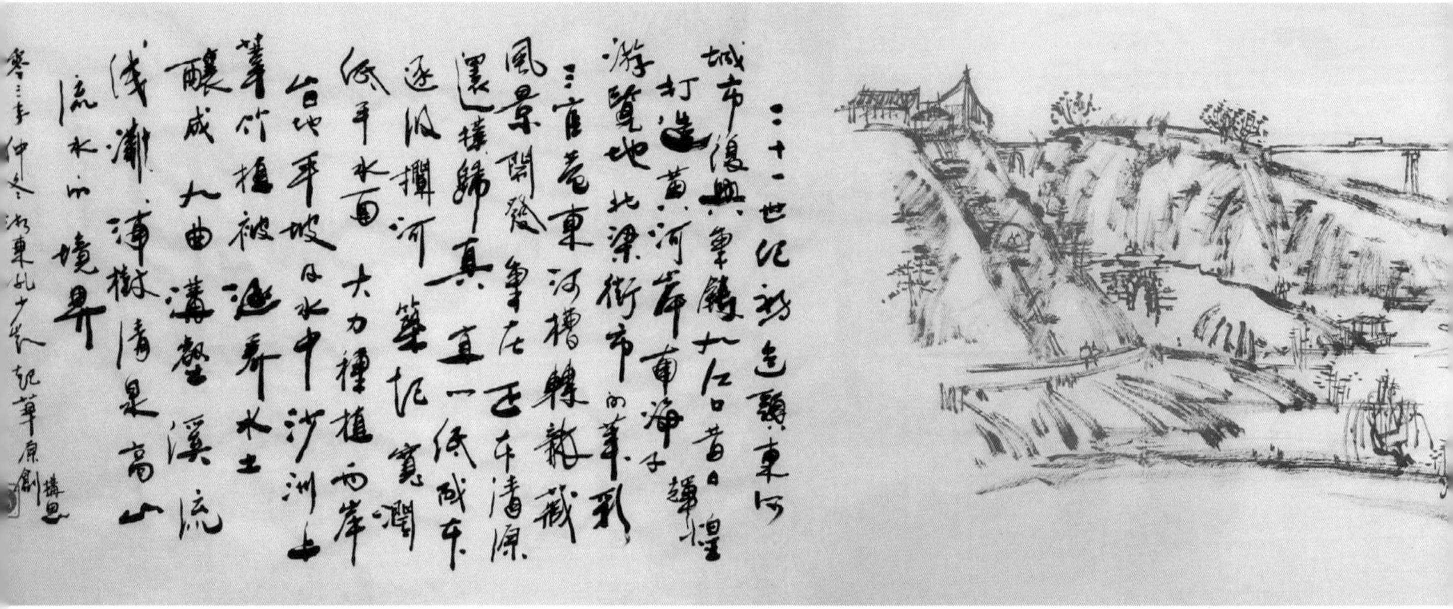

二十一世纪初远眺东河

城市復興筆鶴九合首日輝煌打造黄河岸南海子游覽地北梁街市的筆耕三官蒼東河槽轅龍藏風景開發筆左正本清源還護歸真筆托寬潤一線咸東逐波攔河筆龍寬潤綠平水首大力種植西峰台地平坡瓜永中沙洲土筆竹植被迎弄水土釀成大曲溝壑溪流浅灘蒲樹清泉高山流水的境界

零三年仲冬務東九十長祀草原創撰恩

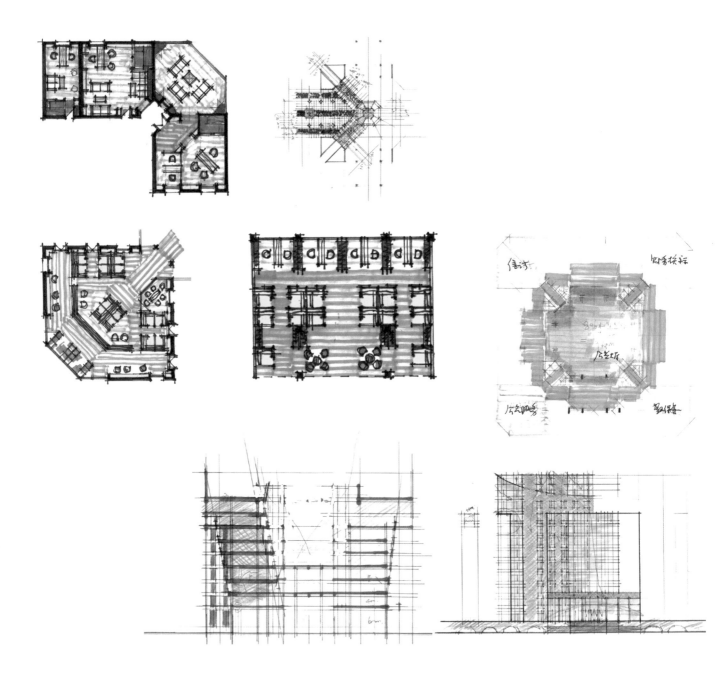

保护自然　保护历史
珍惜　每一寸国土环境
创造建筑城市风景精品

包头东河行政办公楼建筑
方案设计 2005
The Scheme Design of East River Administration Mansion in Baotou: to character a digestive organ plane of office building.
以理性思维建立人流、功能平面与竖向设计的数学模型，致力于打造表现公众兴趣与利益的广厦与聚会中心：方正师石，公平透明。正方器字平面中心部位为公共大厅、多功能厅与屋顶花园，四正方形突出角部因私密性与最佳风景作为负责人办公单元，角部之间为开放式办公区及会议室，角部与中心之间设楼梯电梯厅与洗手间。
四建筑立面，正中部分为48m见方深红色砂岩外墙、加左右8m宽12层通高玻璃幕墙，坐落在第二层主入口平台上。砂岩外墙切出后退8m、40m高32m内凹大门洞，正中两跨为从8m标高开始向外斜挑的玻璃幕墙；四突出角部玻璃幕墙做40m高45°通高切角。

龙藏新城九江口控制性规划
2003.8 - 2004.2
大器建筑设计咨询(上海)有限公司
包头市　城市　规划　设计　研究院

S. Pietro, P. Gallo, *New Italian Houses*, Milano 2006, P12.

S. Pietro, M. Vercelloni, *New American Houses 2,* Milano 2001, P188.

S. Pietro, P. Gallo, *New Villas 2 in Italy*, Milano 2000, P41.

S. Pietro, P. Gallo, *New Villas 2 in Italy*, Milano 2000, P10.

保护自然　保护历史
珍惜　每一寸国土环境
创造建筑城市风景精品

**传统晋风民居四合院优化
发展原创构思 2004**

The Concept to Upgrade Traditional Jin-Style Siheyuan House: removing south room, to add a living room – a meeting & exhibition place to unite lobby, eastern & western rooms.

以前沿策划理念观察传统晋风四合院民居：古代家族聚居，自拱门照壁转折登进院、三跪九拜登堂入室到北厅面见主人；对于现代人而言独缺一个开门见山聚会交流中心。"历史无价"，一份经历两年艰苦思索的四合院优化原创构思：拆除四合院南房，改为露天庭院与西南角的车库；打造从原南房北墙位置向北6m进深拥有南北玻璃落地长窗的客厅，将东面照壁玄关、东厢房、西厢房厨房餐厅融为一体；构成人们一进屋就投入其间一个欢聚交流的雀巢燕窝。

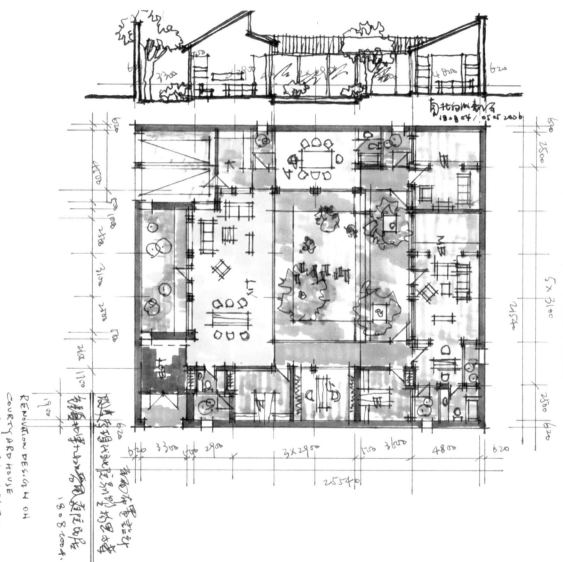

龙藏新城九江口控制性规划
2003.8 – 2004.2
大器建筑设计咨询（上海）有限公司
包头市　城市　规划　设计　研究院

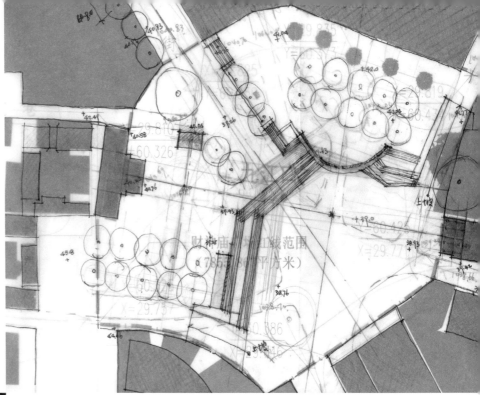

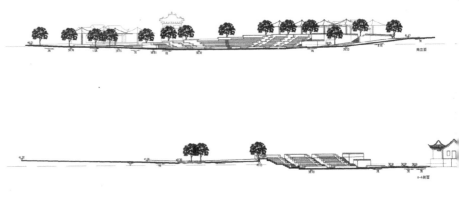
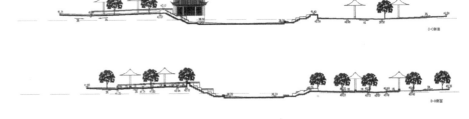

2000年九江口广场重建原构思溯源影像 P198左上1929年斯坦因无声电影财神庙钟楼 左下改建前风雪古戏台
右下改建前风雪财神庙 右上改建前古戏台木构架，彩画及装饰细部 P199 2005年九江口城市广场重建竣工风景

保护自然　　保护历史
珍惜　每一寸国土环境
创造建筑城市风景精品

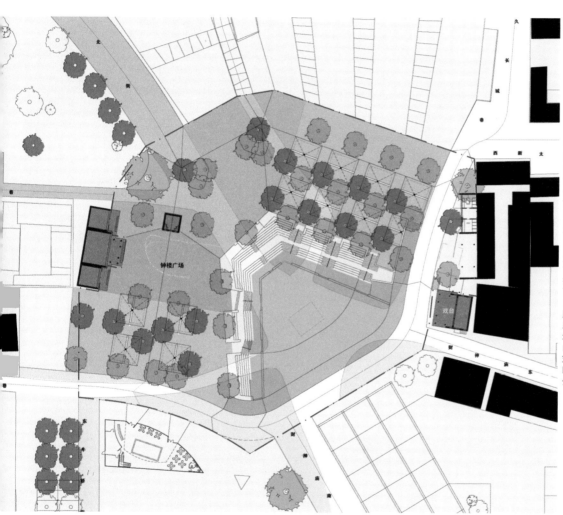

北梁两百年九江口财神庙
城市广场重建设计 2005
The Design of Jiujiangkou Caishenmiao Plaza in Baotou: to preserve & to upgrade 200-year temple & stage pavilion, to open outdoor performance square with natural fountain.

以人为本，大面积利用廉价工料，含财神庙、戏台、露天剧场及东北与西南两个休闲帐篷平台四个功能区。200年财神庙、戏台古建筑修旧如旧，增建公共设施。财神庙山门前按照1927年斯坦因无声电影场景抽象夯原砖墙加钛金属板饰面屋面钟楼。黄木纹铺地广场地面顺应由东北向西南倾斜的天然地势进行雕塑，露天剧场筑东北与西南两组跌落式观众席平台和花池，围合占地1 000m² 梯形平面的小广场，引进山泉凑成一泓明镜浅水池成为儿童戏水的宝地，演出时浅水中木制临时舞台，再现鲁迅小说《社戏》场景。

龙藏新城九江口控制性规划
2003.8 — 2004.2
大器建筑设计咨询（上海）有限公司
包头市　城市　规划　设计　研究院

参考文献 References

冯纪忠.建筑弦柱.上海：科学技术出版社，2003.
冯纪忠.组景刍议.上海：同济大学学报，1979,(4).
冯纪忠.空间设计原理.上海：同济大学学报，1964,(2).
陈从周.说园.上海：同济大学出版社，1984.
郭绍虞，王文生.中国历代文论选.上海：
上海古籍出版社，1978.
沈子丞.历代论画名著汇编.北京：文物出版社，1982.
康熙45年初刻本.二十二子.
上海：上海古籍出版社，1986.
光绪浙江初刻本.全唐诗.上海：上海古籍出版社，1986.
吴楚材，吴调侯.古文观止.北京：中华书局，1978.
方述鑫，林小安，常正光等.甲骨金文字典.
成都：巴蜀书社，1993.
丹纳.艺术哲学.傅雷，译.
北京：中国社会科学出版社，1978.
苏姗·桑纳格.艺术问题.滕守尧，朱疆源，译.北京：
中国社会科学出版社，1978.
李泽厚.美的历程.北京：中国社会科学出版社1978.
简·皮亚杰.结构主义.倪连生，王琳，译.
北京：商务印书馆，1996.
怀特海.科学与近代世界.何钦，译.北京：商务印书馆，
1997.
彭加勒.数学与科学：最后的沉思.李醒民，译.
北京：商务印书馆，1997.
卡里瑟斯.我们为什么有文化.陈丰，译.
沈阳：辽宁教育出版社，1998.
库尔特·勒温.拓扑心理学原理.北京：商务印书馆，2005.
艾森 拉塞尔，保罗·弗里嘉.麦肯锡意识.张涛，赵陵，
译.北京：商务印书馆，2002.
斯蒂芬·霍金.时间简史.许明贤，吴忠超，译.
长沙：湖南科技出版社，1996.
斯蒂芬·霍金.果壳中的宇宙.吴忠超，译.
长沙：湖南科技出版社，2001.
孔少凯.山水游赏空间设计初探.建筑师.
1984-85,(19)(21).
孔少凯.石风景环境思索.建筑学报，1987,(5).
孔少凯.武夷梦忆.上海：解放日报 朝花副刊，
1997.04.29,(9).
孔少凯.石宕绝响.上海：新民晚报 夜光杯副刊，
1999.11.15,(20).
孔少凯.海派别墅新潮.上海美术出版社，2001.
孔少凯.现代教育建筑人文内涵探索.上海：时代建筑,
2002,(2)
孔少凯.外滩光大银行改建工程.南京：室内设计，2002,(5).
孔少凯.现代主义之后环境艺术审美的理性思考.上海：
同济大学学报，2005,(3)
孔少凯.城市土地、文脉与空间发展的理性思考.上海：
长江三角洲经济·空间关系学术讨论会论文集，2006,11.

SIMONDS JOHN O. Landscape architecture: shaping of man's natural environment. New York: F.W. Dodge Corporation,1964.
SIMONDS JOHN O. Earthscape: a manual of environmental planning. New York: McGraw-Hill Corporation, 1978.
CHRISTIAN NORBERG-SCHULZ. Genius Loci: towards a phenomenology of Architecture. Milan: Electa, 1979 & New York: Rezzoli International Publications, Inc.1980.
CHRISTIAN NORBERG-SCHULZ. Concept of dwelling: on the way to figurative architecture. Milan: Electa, 1985 & New York: Rezzoli International Publications, 1985.
CHRISTIAN NORBERG-SCHULZ. Architecture: meaning and place. Milan: Electa, 1986 & New York: Rezzoli, 1988.
JAMES STEELE. Architecture today.
London: Phaidon Press Limited, 1997.
HUGH PEARMAN. Contemporary world architecture.
London: Phaidon Press Limited, 1998.
SHERBAN CANTACUZINO. Re-architecture: old buildings + new uses. New York: Abbeville Press Publication, 1989.
RICHARD H PENNER. Conference center planning & design. New York: Watson-Guptill Publication, 1991.
CHARLES W MOORE, WILLIAM J MICHAEL, WILLIAM TUMBULL Jr. The poetics of gardens. Cambridge MA: MIT Press, 1998.
WILLIAM J MICHAEL. The logic of architecture design, computation and cognition. Cambridge MA: MIT Press, 1990.
SHAOKAI KONG. The meaning of architecture – a thinking process//Proceedings of the Second International Symposium on Architectural Interchange in Asia. Japan, 1998: Architectural institute of Japan.
MILDRED FRIEDMAN, MICHAEL SORKIN, FRANK O GEHRY. Architecture + process Gehry Talks. New York: Universe Publishing, 2002.

参考文献 References

BRAD COLLINS, ELIZABETH ZIMMERMANN. Antoine predock architect 2. New York: Ressoli, 1998.

WILLIAM Jr CURTIS. Le Corbusier: idea & forms. London: Phaidon Press Limited, 1998.

BRUCE BROOKS PFEIFFER. Frank Lloyd Wright. Köln: Benedikt Taschen,1994.

DAVID B BROWNLEE, DAVID G De LONG, VINCENT SCULLY. Louis l. Kahn: in the realm of architecture. New York: Rezzoli International Publications Inc.,1991.

PETER BLUNDELL JONES. Hans Scharoun. London: Phaidon Press Limited, 1998.

CARTER WISEMAN. I. M. Pei profile in America architecture. New York: Harry N Abrams Inc., 1987.

FRANCESCO Dal Co. Tadao Ando complete works. London: Phaidon Press Limited, 1995.

PETER BLAKE. The architecture of Arthur Ericson. Landon: Thames & Hudson, 1988.

OSCAR RIERA OJEDA, LUCAS H. GUERRA Moore ruble yudell houses & housing. Massachusetts: Rockport Publishers, 1994.

JOSEPH RYKWERT. Richard Meier architect. New York: Rizzoli International Publications, Inc., 1984.

OSCAR RIERA OJEDA, LUCAS H GUERRA, RICHARD WESTON. Alvar aalto. London: Phaidon Press Ltd, 1998.

CHRISTIAN NORBERG-SCHULZ, WARREN A JAMES. KPF architecture & urbanism 1986-1992. New York: Rezzoli, 1993.

PAUL GOLDBERGER. KPF architecture& urbanism 1976-1986. New York: Rezzoli, 1991..

PETER BUNCHANAN. Reno Piano building workshop. London: Phaidon Press Limited, 1998.

PHILIP JODIDIO. Sir Norman Foster. Köln: Benedikt Taschen, 1997.

VINCENT SCULLY, ELIZABETH KRAFT. Robert A M Stern buildings & projects 1987-1992. New York: Rezzoli, 1992.

COLLIN ROWE, PETER ARNDELL, TED BICKFORD. James Stirling buildings & projects. New York: Rizzoli, 1984.

CATHERINE SLESSOR. 100 of the world's best houses. Mulgrave: Image Publishing Group Pty Ltd., 2002.

CLARE MELHUISH. Modern houses 2. London: Phaidon Press Limited, 2000.

RANDALL WHITEHEAD. Residential lighting. Rockport Massachusetts: Rockport Publishers, 1993.

MELANIE SIMO, DDVID DILLON. Sasaki Associates: integrated environments. Washington DC: Spacemarker, 1997.

MARGARET COTTOM-WINSLOW. International landscape design. New York: PBC International Inc., 1991.

PBC. Club & Resorts: designing for recreation and leisure. New York: PBC International Inc., 1993.

PACO ASCENSIO, QUIM RESELL. Waterfront homes. New York: Loft Publishing's and HBL, 2000.

HIDENOBA, JINNAI, MUROTANI, BUNJI. Italian aquascape. Tokyo: Process Architecture Co., Ltd.,1993

ANN BREEN, DICK RIGBY. The new waterfront: a worldwide urban success story. London: Thames & Hudson 1996

CAROL SOUCEK KING. At home & at work: architects' and designers' empowered spaces. New York: PBC International Inc., 1993.

SANFORD KWINTER. Wiseman Architects. Mulgrave: Image Publishing Group Pty ltd., 1995.

MICHAEL CROSBIE. Cesar Pelli. Mulgrave: Image Publishing Group Pty ltd., 1993.

KISHO KUROKAWA. Kisho Kurokawa. Mulgrave: Image Publishing Group Pty ltd., 1995.

RICHARD GUY WILSON. Haztman-Cox. Mulgrave: Image Publishing Group Pty ltd., 1994.

JOAN OCMAN. Skidmore, Owings & Merrill. Mulgrave: Image Publishing Group Pty ltd., 1995.

PHILLIP COX, MICHAEL RAYNER. Cox Architects. Mulgrave: Image Publishing Group Pty ltd., 1994.

CLARE MELHUISH. Tezzy Fazzel. Mulgrave: Image Publishing Group Pty ltd., 1994.

ARUP ASSOCIATES. Arup Associates. Musgrave: Image Publishing Group Pty ltd., 1994.

JAHN. Murphy. Mulgrave: Image Publishing Group Pty ltd., 1995.

JOHN R DALE. Barton Myers. Mulgrave: Image Publishing Group Pty ltd., 1994.

参考文献 References

MASAYUKI FUCHGAMI, MARIO PISANI. Architecture • studio. Mulgrave: Image Publishing ltd., 1995.
KAY KAISER. Richard Keating. Mulgrave: Image Publishing Group Pty ltd., 1996.
LARRY PULL FULLER. RTKL 96: the top fifty years. Mulgrave: Image Publishing Group Pty ltd., 1996.
ERIKA ROSENFELD. NBBT. Mulgrave: Image Publishing Group Pty ltd., 1997.
DENNIS SHARP. Harry Seidler. Mulgrave: Image Publishing Group Pty ltd., 1997.
CHARLES GWATHMEY. Gwathmey Seigei. Mulgrave: Image Publishing Group Pty ltd., 1998.
SILVIO SAN PIETRO, ALESSADRA VASILE. New exhibits 3 made in Italy. Milano: Edizioni L' Archivolto, 2006.
SILVIO SAN PIETRO, MIGLIORE, SERVETTO. New exhibits 2 made in Italy. Milano: Edizioni L' Archivolto, 2000.
SILVIO SAN PIETRO, ALESSADRA VASILE. New offices in Italy. Milano: Edizioni L' Archivolto, 2003.
SILVIO SAN PIETRO, PAOLO GALLO. Lofts 2 in Italy. Milano: Edizioni L' Archivolto, 2003.
SILVIO SAN PIETRO, PAOLO GALLO. Living in Milan. Milano: Edizioni L' Archivolto, 2001.
SILVIO SAN PIETRO, ANNAMARIA SCEVOLA. Urban interior 3 in Italy. Milano: Edizioni L' Archivolto, 2000.
SILVIO SAN PIETRO, PAOLO GALLO. New shops made in Italy. Milano: Edizioni L' Archivolto, 1994.
SILVIO SAN PIETRO, PAOLO GALLO. New shops 5 made in Italy. Milano: Edizioni L' Archivolto, 1998.
SILVIO SAN PIETRO, PAOLO GALLO. New shops 6 made in Italy. Milano: Edizioni L' Archivolto, 2000.
SILVIO SAN PIETRO, PAOLO GALLO. New shops 8 made in Italy. Milano: Edizioni L' Archivolto, 2005.
SILVIO SAN PIETRO, PAOLO GALLO. Renovated houses. Milano: Edizioni L' Archivolto, 1999.
SILVIO SAN PIETRO, PAOLO GALLO. New villas 2 in Italy. Milano: Edizioni L' Archivolto, 2000.
SILVIO SAN PIETRO, PAOLO GALLO. New Italian houses. Milano: Edizioni L' Archivolto, 2006.
SILVIO SAN PIETRO, METTE VERCELLONI. New offices in USA. Milano: Edizioni L' Archivolto, 1998.
SILVIO SAN PIETRO, METTE VERCELLONI. New stores in USA. Milano: Edizioni L' Archivolto, 1999.
SILVIO SAN PIETRO, METTE VERCELLONI. New showrooms & galleries in USA. Milano: Edizioni L' Archivolto, 1999.
SILVIO SAN PIETRO, METTE VERCELLONI. New american houses. Milano: Edizioni L' Archivolto, 2001.

Architecture & Photograph Credits

Humboldt Bibliothek, Berlin Germany: Crossroad School, Santa Monica California: Science Complex, University of Oregon, Eugene and California Center for Arts, Escondido:
design - Moore Ruble Yudell Architects & Planners, photos - Timothy Hursley
Seabird Island School, British Columbia Canada: design & photos - Patkau Architects Inc.
Hysolar Institute, Stuttgart University Germany: design & photos - Behnisch & Partner Architects
Isala College, Silvode Netherlands: design & photos - Mecanno Architects
Cite' Scolaire, Lyons France: design -Jourda & Perraudin
Centenary Building, University of Solford England: design - Hodder Associates
Jewish School in Berlin Germany: design – Zvi Hecker Architect, photos - Michael Kruger
Ventana Vista Elementary School, Tucson Arizona: design - Antonie Predock Architect, photos - Timothy Hursley
Dance Facility, University of California San Diego: design - Antonie Predock Architect, photos - Anne Garrison

后记

　　人，只是沧海一粟，人生，仅仅留下一丝过眼烟云。而当其全身心投入到学习之中，就化为芸芸众生历史潮流中闪烁波光天色的点点滴滴。

　　《建筑体验》与《设计原创》两本书，是追随老师与同学们如砌如磋如琢如磨的雕塑。愿其成为对年青学子们的小小奉献。

　　对于从经典书籍文献中汲取的人类思想与经验结晶，谨向中外作者与出版社致以最诚挚的感谢。

浙东　孔少凯　2010新年写于海上

Postscript

　　Man has only been a drop in an ocean. Man's life has been a tiny bit of a transient fleeting clouds. When all of a man's body & mind are thrown into his study, it will be sublimated to the shining light of ripple & sky within history flows.

　　Two books *Experiencing Architecture Designs* & *The Designs for Poetic Dwelling*, are the sculpture which has been reated fallowing professors' instructions and with classmates' common endeavor. I hope it be a contribution.

　　For the crystals of both concept & experience derived from the books & the documents, I would express here the most sincerely thanks & respects to the authors & the publishing houses in China & in the world.

作者背景

　　自幼习竹笛山水，孔少凯先生10岁被保送到浙江省重点绍兴中学，13岁学萃取重铀酸铵实验，立志专攻数理天文报国。1964年全国高考宁波地区第一名，"文革"十年在新疆从事建筑木工、木模工、文艺演奏与创作，1976年脑外伤抢救手术后投身建筑学。1980-1988年师从同济大学冯纪忠教授攻读风景建筑硕士并任教，1988-1992年师从马尼托巴大学莱特教授与路易斯·康弟子纳尔逊教授攻读城市设计硕士，1992年归国任职香港何弢设计所，后主持个人工作室。外祖父孔墉为辛亥革命家、教育家、抗日阵亡将领与革命烈士，孔少凯先生2005年出任复旦大学上海视觉艺术学院客座教授，专注于建筑体验与原创设计的研究、教学与交流。

Auther Background

　　Studying flute & Chinese painting from childhood, and when 10 years old recommended studying in Shaoxing High School where Prof. Cai Yuanpei was the principle 1898, Mr.Kong Shaokai has been intended to study math & astronomy and placed No.1 among the ninbo region in 1964's national examination for university study. During the Cultural Revolution worked as construction carpenter, molder, instrument player and writer, he has devoted himself into architecture field from 1976 when received a head injury saving operation.

　　Studying for M.Arch & teaching in U of Tongji under the direction of distinguish Prof. Feng Jizhong 1980-1988, studying for M.LA & guest teaching in U of Manitoba under the direction of distinguish Prof. Alexander E. Rattray & Prof. Carl R. Nelson Jr.1988-1992, he returned his motherland 1992, working as senior architect in Dr. Taoho's office for five years, and in charge of a personal design studio later.

　　Focused into the academic research, teaching and exchange on *Experiencing Architecture Designs & The Designs for Poetic Dwelling*, Kong Shaokai taken up the post of guest professor in SIVA Fudan University 2005. His grandfather in law had been a Xinhai revolutionist, educationist, Anti-Japanese sacrificed general and martyr.

图书在版编目（CIP）数据

设计原创NO.2：汉英对照 /孔少凯编撰. —天津：天津大学出版社，2013.1
　ISBN 978-7-5618-4409-0

　Ⅰ.①设… Ⅱ.①孔… Ⅲ.①建筑设计—研究—汉、英 Ⅳ.①TU2

中国版本图书馆CIP数据核字(2012)第186199号

责任编辑　高亚洲
装帧设计　王　姣

出版发行	天津大学出版社
出　版　人	杨欢
地　　　址	天津市卫津路92号天津大学内（邮编：300072）
电　　　话	发行部：022 – 27403647　邮购部：022 – 27402742
网　　　址	www.publish.tju.edu.cn
印　　　刷	上海锦良印刷厂
经　　　销	全国各地新华书店
开　　　本	210mm × 210mm
印　　　张	13.5
字　　　数	157千
版　　　次	2013年1月第1版
印　　　次	2013年1月第1次
定　　　价	79.00元